Perspectives in Inflammation Biology

Ena Ray Banerjee

Perspectives in Inflammation Biology

Ena Ray Banerjee
Immunology & Regenerative
Medicine Research Unit
University College of Science, Technology
and Agriculture
Kolkata, India

ISBN 978-81-322-1577-6 ISBN 978-81-322-1578-3 (eBook)
DOI 10.1007/978-81-322-1578-3
Springer New Delhi Heidelberg New York Dordrecht London

Library of Congress Control Number: 2013951404

Printed on acid-free paper

Springer is part of Springer Science+Business Media (www.springer.com)

To my children Urbi, Adit, and Arit

Foreword

As a researcher in basic and translational Immunobiology and Drug Discovery and later on, hyper-specializing into lung inflammation and lung stem cell engineering, I have felt, on the one hand as an academician deficit of a text book that gives the beginner, an *in depth* grasp of the subject and its nuances and on the other hand, a comprehensive *ready reckoner* for the bench scientist, that offers insight and troubleshooting ability keeping in mind the nitty gritties of the techniques and tools necessary for investigation. Hence, the endeavour for this discourse, which shall attempt on a four pronged approach to the same basic tools and techniques to alleviate human suffering, viz. firstly introduction to the basic biological cum clinical and drug discovery aspects of pulmonary and systemic inflammation; followed by detailed discussion on preclinical models of pulmonary and systemic inflammation researcher both in academia and the drug discovery industry; suceeded by the description of studies done on the roles of some key molecules in acute allergic asthma under which focuses on two major areas- research area 1 and 2 which talk respectively about exploring the roles of enantiomers of albuterol, the OTC drug of choice for acute asthma management and studies on prophylactic and therapeutic strategies to combat some local and systemic inflammatory pathologies under which chapter separate sections or sub-chapters are dedicated to investigating the roles of integrins $\alpha 4$ and $\beta 2$, the three selectins, gp91phox and MMP-12 in acute allergic asthma. This section precedes studies on the roles of integrins $\alpha 4$ and $\beta 2$ in chronic allergic asthma and in lymphopoiesis & homing and the last chapter is dedicated to the study of their roles in aseptic peritonintis. The information, mostly trouble-shooting tips from a conceptual as well as practical standpoint, I have often found, either vague or altogether lacking in research articles or text books, but have always felt to be critical for a successful reproduction of a previous work or a continuation of an ongoing problem. I sincerely hope that this is an honest and thorough work and our effort to share the finer details shall benefit the basic in academics as well as the translational researcher in the industry in their quest as well as give students, beginner and senior, a glimpse into possibilities of the excitement that science has to offer and to choose their career carefully, something that is often beyond the scope of a standard text book and perhaps often lost in translation in the technical jargon of an erudite journal. Our research has indeed filled lacuna in the existing literature in the relevant fields, now we hope that this discourse shall excite future students and researchers and initiate them into the art of science. Each chapter highlights recent advances in a selected domain in lung disease research. This book shall serve as a comprehensive resource for both scientists and clinicians studying various aspects of fundamental and translational health sciences and medicine and shall provide a single reference comprising both basic and specialized information.

Kolkata, India Ena Ray Banerjee

Preface

At the beginning of my research career, when I was toying with the choice of various areas of possible study, immunology had appealed to me the most, to my mind at that time, when one is usually faced with the dilemma of a choice of career that offers a quick entry into the world of professionals versus a slightly longer-drawn-out yet infinitely rewarding career of an esoteric academician. Unfortunately, I did not get many role models to follow nor very sound career advice post my completion of M.Sc. in Zoology from the University of Calcutta. The options were either applied fields strictly in Zoology, such as the much in demand fisheries or entomology, or the slightly archaic career in museology and taxonomy or at best genetics which was also rather for a basic research career, with very little possibilities of translation. In due course, I qualified the UGC NET exam and opted for research in immunobiology, steering clear of infectious diseases and choosing instead cytokine biology, and joined Indian Institute of Chemical Biology, Calcutta, and worked on immunomodulation of interleukin 8 receptor expression on human polymorphonuclear neutrophils using various immunomodulators such as PMA and LPS. My work, while exposing me to the finer techniques and tools of our trade, also gave me an insight into the possibilities of research traversing various areas of immunobiology using rather repetitive yet useful assays. In fact, the take-home message here would be that receptor biology is an all-encompassing field that can be potentially applied to one of many receptor types and should help the researcher explore receptors pertinent to various diseases with even limited exposure to tools and techniques. Once you get a feel of the subject, you are initiated, so to speak, it becomes easy to use the basics and extrapolate into realms that can foray into very diverse fields of biological and related sciences.

My Ph.D. research was completed in roughly three years (1994–1997) with some fairly good publications in national and international journals, and having presented my work in some relevant conferences, I received my degree from Jadavpur University the next year as soon as my first child, a daughter, Urbi was born. Subsequently, I was told most solemnly by my mother, an East–west fellow and an accomplished anthropologist herself, that an academic career was the best career option for a woman. I therefore applied for and was duly selected for lectureship by College Service Commission and Public Service Commission as well and started teaching undergraduate Zoology which I did for 8 years in total. After my son Adit was born, we left for a postdoctoral stint in the USA where I worked on hematology, exploring the roles of various integrins and selectins in inflammation of acute allergic asthma using various genetic knockout models of mice. From work on lung inflammation, I naturally progressed to use of cell-based therapy to ameliorate degenerative lung diseases, at the base of which lies inflammation. This took me into the realm of stem cells and regenerative medicine, and I became interested in exploring stem cell niches in the lung. We then returned home, and my entry into the world of applied biological research began with my foray into the Biotech and pharmaceutical industry where I briefly headed research with preclinical models of lung inflammation and aiming various libraries of chemical moieties toward various targets. Returning to academics after a couple of years (because the pressures and constraints of the industry are not always pleasant for us free-thinking academicians), my lab was initiated into working on the areas best identified as relevant in my experience as a university teacher and researcher as well as my exposure as an industry work force and that I felt I was best trained to do, and these are the following: (a) inflammation biology being my initiation into the world of research still remained my strong point as I understood how receptors and ligands interacted and what that meant for the overall signaling of the cell. In this I chose inflammation and drug discovery in inflammatory diseases as being the most relevant as far as translation is concerned, not, however, compromising basic research that asked fundamental questions. We work on the following primary models of inflammation where we ask questions on disease markers as well as test and screen drugs on them: allergen-induced acute and chronic allergic asthma, aseptic peritonitis, noninfectious colitis, idiopathic pulmonary fibrosis, and various transplantation models. In line with my training in Immunology, we also work on developing a platform technology for novel format camelid antibodies of diverse antigen: disease markers and cell differentiation markers. (b) Stem cell and regenerative medicine is the other area where I naturally migrated into while working with the lung as my primary target tissue for therapy. While I was trying to understand the hematological parameters and roles of various integrins in the onset of allergic inflammation as well as lymphoid homing in diseased state in the lung, regenerative medicine was the obvious transition where pharmacological intervention was not enough. So we began exploring pathways to convert embryonic stem cells with pluripotent characters into lung cells, a world of possibilities opened before us, and we began using various avenues such as endodermal differentiation, with or without embryoid body formation, varying density and giving heat shocks and co-culture with other cells, gently engineering the pluripotent cells into lineage of our choice, namely, nonciliated alveolar epithelial tissue. In due course, validation was done with transplantation models where homing and engraftment were studied.

So here is hoping that many more curious souls like me shall enter the world of research asking questions and opening new vistas for generations to come and enjoy the thrill of science all the way.

August 15, 2012 Ena Ray Banerjee
Kolkata, India

Acknowledgements

I would like to take this opportunity to thank the people who were instrumental in helping me write this book. The first is Dr. Richa Sharma who was on a scouting mission in Calcutta University and first floated the idea to me. While there have been many invitations to be part of books or direct offer for a commission, what with the children and their demands and other impediments, it never took off. So I thank Richa for her follow-up on this and here I am finally sitting down to write the book. My infant son, Arit, was in the hospital fighting for his life when I first started working on this book, and while he slept, I kept my anxiety at bay by channeling my energies into thinking and formulating this monograph. It helped keep me sane in the face of the greatest crisis of my life when my baby was at death's door and valiantly fighting back. I thank my son and his guardian angels for flagging off what I hope will be a helpful addendum in the repertoire of world research in Inflammation Biology and Pulmonary and Critical Care Medicine and help rapid bench-to-bedside translation of knowledge gained from work such as outlined in the following pages to actually help people in need. It will be incomplete if I do not acknowledge my daughter Urbi and my elder son Adit who have always taken pride in what their mother does and has unflinching faith in her work, both at home and at work. I hope when all three of my children grow up and are able to understand the field better, they will appreciate the contents of this work. I shall also be amiss if I do not mention my parents and their sacrifice to take care of me and sometimes, through their superciliousness, push me to challenge myself to do better. Last but not the least, I acknowledge Dr. Umesh Singh, whose contribution to my well-being and support are unparalleled. I am truly blessed to have had the good fortune to share my life with all these lovely people, my family, friends, colleagues, and students who all inspire me in some way or the other every moment of every day.

February 12, 2013
Kolkata, India

Contents

1 Pulmonary and Systemic Inflammation 1
Introduction 1
Pulmonary Models of Inflammation 2
What Is Inflammation? 2
Acute and Chronic Inflammation 2
Asthma and COPD 3
Key Role of Inflammation in Respiratory and Systemic Immune Disorders 3
Unmet Needs in COPD/Asthma Therapy 4
References 6

2 Preclinical Models of Acute and Chronic Models of Lung Inflammation 7
Animals for Study of Role of α4 Integrin 7
The α4 Integrin Knockout Mouse 7
The β2 Integrin Knockout Mouse or CD18 Knockout Mouse 8
Targeting Construct and Generation of Mutant Mice 8
Study Design for the Development of Murine Chronic Allergic Asthma Model 15
Fluorescein-Activated Cell Sorter (FACS) Analysis 16
Study Design to Identify Resident Stem Cells of the Lung 17

3 Studying the Roles of Some Key Molecules in Acute Allergic Asthma 19
Research Area 1: Enantiomers of Albuterol, the OTC Drug of Choice for Acute Asthma Management 19
Background and Relevance of the Study 19
Results in a Nutshell 20
Detailed Results 20
Discussion 25
Conclusion 28
Materials and Methods Used in the Study 28
Research Area 2. Studies on Prophylactic and Therapeutic Strategies to Combat Some Local and Systemic Inflammatory Pathologies 30
Overall Objective of This Series of Studies 30

Subchapter 1: Role of Integrin α4 (VLA – Very Late Antigen 4) and Integrin β2 (CD18) in a Pulmonary Inflammatory and a Systemic Disease Model Using Genetic Knockout Mice 30
Summary of the Study 30
Background and Objective of the Study 31
Methods 32
Results in a Nutshell 33
Detailed Results 33
Discussion 41
Conclusions 44
Subchapter 2: Role of E-, L-, and P-Selectins in the Onset, Maintenance, and Development of Acute Allergic Asthma 44
Summary of the Study 44
Background and Scope of the Study 44
Results in a Nutshell 46
Results 46
Discussion 51
Conclusion 57
Materials and Methods 57
Subchapter 3: Role of gp91phox Subunit of NADPH Oxidase and MMP-12 in an Acute Inflammatory and an Acute Degenerative Pulmonary Disease Model Using Genetic Knockout Mice 60
Summary of the Study 60
Introduction 60
Materials and Methods 63
Results 66
Discussion 78
References 85

4 Role of Integrins α4 and β2 Onset and Development of Chronic Allergic Asthma in Mice 91
Background and Objective of the Research Undertaken 92
Results in a Nutshell 92
Detailed Results 93
α4Δ/Δ Mice Fail to Develop AHR to Chronic Airway Challenge by Allergen 93
Migration of Leukocytes from Circulation to Lung and to Airways 93
Inflammation and Fibrosis in the Lungs in Response to Chronic OVA Challenge 93
Th2/Th1 Cytokines in BALf and Plasma and IgE and IgG1 Levels in Plasma 95
Soluble VCAM-1 in BALf and Plasma and VCAM-1 Expression in the Lung 95
TGF-β1 and Soluble Collagen in BALf 96
Discussion 97
Conclusion 105

Materials and Methods 105
Animals 105
Induction of Chronic Allergic Asthma 105
Bronchoalveolar Lavage Fluid 105
Lung Parenchyma Cell Recovery 105
Lung Histology 106
Lung Immunohistochemical Staining 106
Fluorescein-Activated Cell Sorter (FACS) Analysis 106
Cytokines 107
OVA-Specific IgE and IgG1 in Plasma 107
Pulmonary Fibrosis 107
Soluble VCAM-1 and Soluble Collagen in Lung Homogenate . . 107
Lung Function Testing 107
Th2 Differentiation, Intracellular Staining, and ELISA Assay for IL-17A and IFN-γ 107
Statistics 108
References 108

5 Role of Integrin $\alpha 4$ (VLA – Very Late Antigen 4) in Lymphopoiesis by Short- and Long-Term Transplantation Studies in Genetic Knockout Model of Mice 111
Introduction 112
Materials and Methods 113
Mice 113
Antibodies and Fluorescein-Activated Cell-Sorting (FACS) Evaluation 113
Preparation of Tissues for Cellularity and FACS Evaluation 113
Immunohistochemistry 114
Immunization with Trinitrophenyl Ovalbumin (TNP-OVA) 114
Proliferative Responses and Cytokine Secretions by Lymphoid Cells 114
Results 114
Hemopoietic Reconstitution by Donor Cells in Rag 2–/– Recipients 114
Repopulation of Lymphoid Organs with $\alpha 4\Delta/\Delta$ Donor Cells . . . 118
Thymus 118
Peripheral Lymph Nodes 118
PPs and MLNs 119
Functional Status of α4-Deficient Lymphoid Cells 120
T Cells 120
Discussion 121
Conclusion 125
References 126

6 Studying the Roles of Some Critical Molecules in Systemic Inflammation 129
Introduction 130
Materials and Methods 130
Mice 130

Antibodies 131
Peritoneal Inflammation 131
Fluorescein-Activated Cell-Sorting Analysis 131
Actin Polymerization 131
Ca^2+ Mobilization 132
Statistical Analysis 132
Results 132
Animal Models 132
To Study Unique and Redundant Roles of α4 and β2 Integrins . . 132
Kinetics of Migration of Various Leukocyte Subsets In Vivo . . . 132
Discussion 140
Conclusions 141
References 141

Highlights of the Important Findings from the Critical Analyses of the Data 145

About the Author 147

1 Pulmonary and Systemic Inflammation

Introduction

In order to begin work on a disease, with the aim to ameliorate human suffering, a researcher must first educate herself about the nitty-gritties of what is involved and what is at stake. This involves recreating the inner workings of a human disease in a nonhuman but closely related animal so as to facilitate delineation of each and every fine detail of the onset, etiology, establishment, development, progression, and exacerbations (cyclical manifestations in an exaggerated form of the disease pathology). In order to understand these inner workings, the model has to be simple enough and yet similar enough to be of any use to alleviate suffering of the human patients. The bench-to-bedside strategy therefore aims to recreate a complex disease first, part by part, outside the body (ex vivo) which enables easy dissection of nodal points of disease onset-progression, and then the entire composite disease phenotype is addressed using a whole organism by various treatments such as administration of a molecule (chemically induced), be it a drug, a polymer, an antibody or a peptide, or even an allergen or toxin, by various routes of administration, which, over various time intervals, usually manifest in a complex human disease in the subhuman primate or non-primate.

The first criterion therefore, for a researcher of biomedicine, is the choice of the animal once the disease phenotype she is particularly interested in is chosen in its finer nuances. The choice of the animal and the strain is important because there are some that are refractory to the treatment and do not satisfactorily express the disease phenotype and some that moderately express the disease. In both cases, it is a waste of time since the data from the control animal will not appreciably differ from that of the treated. Review on such strains or original research articles on this topic comparing and contrasting how disease manifestation differs depending on strain and treatment routes are very valuable. We shall discuss the same in detail for the disease models we are particularly interested in.

The next yardstick in choosing the model will be the readouts based on subtle variations in treatment. For example, brief initial adjuvant-aided sensitization and follow-up that prolonged local challenge in an allergic model is likely to invoke a cytokine-mediated pathway more powerfully than if slightly prolonged sensitization initiates the model and local challenges are somewhat curtailed (to half or two third of the earlier regimen) when a more pronounced B-cell-IgE-directed allergic response is seen. This also will be addressed in the following chapter.

The overall assessment of the "satisfactoriness" of the disease shall be based on the following features:

(a) That the disease expression is significantly distinguishable from the negative control or placebo-treated subject
(b) That the set criteria for simulation with the human counterpart are appreciably quantifiable

E.R. Banerjee, *Perspectives in Inflammation Biology*,
DOI 10.1007/978-81-322-1578-3_1, © Springer India 2014

(c) That the reversal of the disease (posttreatment intervention of a known pathway) is easily detectable (either by objective quantitative evaluation or by blinded qualitative estimation)

To fulfill these checkpoints, a disease model has to be extremely well characterized so that it is easy to interpret should there be even subtle shifts posttreatment. Well-characterized genetic knockout models of mice are therefore preferred for models of diseases as information on key regulatory molecules are already unambiguously available in the public domain.

Our work mainly involves pulmonary and systemic models of inflammatory diseases. We also work on disease models of degenerative diseases (will be discussed in detail in Chap. 2 of this book). Suffice it to say that most diseases have both inflammatory and degenerative components and while one scientist may be interested to explore the inflammatory component of it, another may be interested to study regeneration. Keeping this in mind, we shall endeavor to share in detail the models of interest to us and those that are used regularly in our laboratory and those, which may we modestly say, "have a handle on!"

Pulmonary Models of Inflammation

What Is Inflammation?

Inflammation may be defined as the sequential chain of events that herald the clearance of pathogens, rogue cells, and insult to the homeostasis of the system and is characterized by complex biological and biochemical response mainly via blood-borne inflammatory cells and the soluble mediators. Receptors and their interaction with these ligands orchestrate the complicated tango of cells that stage the drama of inflammation. In the absence of inflammation, an infection would never heal. This however necessitates the correct and timely control of the phenomenon; otherwise, loss of this master control by Inflammation Regulatory Elements (IRE), the drama becomes a saga of chronic and cyclical exacerbation leading ultimately to tissue degeneration and consequent structural changes. While pharmaceutical intervention may interfere with inflammation locally, systemic inflammation has to be managed more carefully as this is also a vital requirement by the body and any consequent break in signaling may harm the entire homeostasis. The challenge therefore for the researcher is to seek ways to limit the field of operation by anti-inflammatory agents.

It is a complex biological response of vascular tissues to appropriate stimuli in which there is constant and ever-changing interplay among cells of the circulation, local resident cells, soluble mediators, and genetic factors that form a myriad of signaling networks. The role players in specific diseases change as do their particular contribution to the response to the specific pathogen, namely, a pathogen, a damaged cell, or an irritant such as an allergen. It is a protective attempt by the organism to remove the insult and initiate the healing process. So this is an essential phenomenon. However, if it were to run unchecked, it would jeopardize the survival of the organism itself and therefore need careful monitoring and therein lies the importance of the study of the pathways regulating the initiation, establishment, and progression of pathophysiology, and the resulting information generated may be used as weapons to counter and control such unchecked inflammation during the development of a disease.

Acute and Chronic Inflammation

Inflammation can be classified as either *acute* or *chronic*. *Acute inflammation* is the initial response of the body to harmful stimuli and is achieved by the increased movement of *plasma* and *leukocytes* from the blood into the injured tissues. A cascade of biochemical events propagates and matures the inflammatory response, involving the local *vascular system*, the *immune system*, and various cells within the injured tissue. Prolonged inflammation, known as *chronic inflammation*, leads to a progressive shift in the type of cells which are present at the site of inflammation and is characterized by simultaneous destruction and healing of the tissue from the inflammatory process.

Acute inflammation is a short-term process which is characterized by the classic signs of inflammation – swelling, redness, pain, heat, and loss of function – due to the infiltration of the tissues by plasma and *leukocytes*. It occurs as long as the injurious stimulus is present and ceases once the stimulus has been removed, broken down, or walled off by scarring (*fibrosis*). The process of acute inflammation is initiated by the blood vessels local to the injured tissue, which alter to allow the exudation of *plasma* proteins and *leukocytes* into the surrounding tissue. The increased flow of fluid into the tissue causes the characteristic swelling associated with inflammation, and the increased blood flow to the area causes the reddened color and increased heat. The blood vessels also alter to permit the extravasation of leukocytes through the *endothelium* and *basement membrane* constituting the blood vessel. Once in the tissue, the cells migrate along a *chemotactic* gradient to reach the site of injury, where they can attempt to remove the stimulus and repair the tissue.

Chronic inflammation is a pathological condition characterized by concurrent active inflammation, tissue destruction, and attempts at repair. Chronic inflammation is not characterized by the classic signs of acute inflammation listed above. Instead, chronically inflamed tissue is characterized by the infiltration of mononuclear immune cells (*monocytes*, *macrophages*, *lymphocytes*, and *plasma cells*), tissue destruction, and attempts at healing, which include *angiogenesis* and *fibrosis*.

Asthma and COPD

Asthma and COPD are chronic conditions that take an enormous toll on patients, healthcare providers, and society. In the context of disease management, acute exacerbations are important clinical events in both illnesses that largely contribute to an increase in mortality and morbidity. Although these diseases are treated with the same drugs, they differ significantly in their underlying etiology. The underlying characteristics of both conditions however involve inflammatory changes in the respiratory tract, while the specific nature and the reversibility of these processes largely differ in each entity and disease stage. Both are characterized by lung inflammation; however, patients with asthma suffer largely from reversible airflow obstruction, whereas patients with COPD experience a continuous decline in lung function as disease progresses.

Asthma and chronic obstructive pulmonary disease (COPD) together form the third leading cause of death in both developed and developing countries, and annual direct and indirect cost of healthcare is more than $50 billion in the USA alone. It is estimated that there were about 45 million patients with asthma in the seven major markets in 2006, with a stabilizing prevalence. These inflammatory disorders are increasing in prevalence, and while most asthmatic patients respond well to current therapies, a small percent of nonresponders (10 %) account for greater than 50 % of healthcare costs. By 2020, India alone will account for 18 % of the 8.4 million tobacco-related deaths globally [1]. In China, COPD is one of the high frequency causes of death followed closely by ischemic heart disease and cardiovascular disease [2].

Key Role of Inflammation in Respiratory and Systemic Immune Disorders

Inflammation is key to etiology of most respiratory disorders, and while it is critical for the body's defense against infections and tissue damage, it has increasingly become clear that there is a fine balance between the beneficial effects of inflammation cascades and potential for tissue destruction in the long term. If they are not controlled or resolved, inflammation cascades lead to development of diseases such as chronic asthma, rheumatoid arthritis, psoriasis, multiple sclerosis, and inflammatory bowel disease. The specific characteristics of inflammatory response in each disease and site of inflammation may differ, but recruitment and activation of inflammatory cells and changes in structural cells remain a universal feature. This is associated with a concomitant increase in the expression of components of inflammatory cascade including cytokines, chemokines, growth factors, enzymes, receptors, adhesion molecules, and other biochemical mediators.

Asthma is characterized by complexity resulting from the interactions among a variety of biomechanical, immunological, and biochemical processes that lead to airway narrowing. The pathogenesis of allergic asthma involves the recruitment and activation of many inflammatory and structural cells, all of which release mediators that result in typical pathological changes of asthma. The chronic airway inflammation of asthma is unique in that the airway wall is infiltrated by T lymphocytes of the T-helper (Th) type 2 phenotype, eosinophils, macrophages/monocytes, and mast cells. Accumulation of inflammatory cells in the lung and airways, epithelial desquamation, goblet cell hyperplasia, mucus hypersecretion, and thickening of submucosa resulting in bronchoconstriction and airway hyperresponsiveness are important features of asthma [3]. Both cells from among the circulating leukocytes such as Th2 lymphocytes, mature plasma cells expressing IgE, eosinophils [3], and neutrophils as well as local resident and structural cells constituting the "respiratory membrane" (airway epithelial cells, fibroblasts, resident macrophages, bronchial smooth muscle cells, mast cells, etc.) contribute to the pathogenesis of asthma [4]. Airway hyperresponsiveness of asthma is clinically associated with recurrent episodes of wheezing, breathlessness, chest tightness, and coughing, particularly at night or in early morning. These episodes are associated with widespread but variable airflow obstruction that is often reversible either spontaneously or with treatment. Furthermore, during exacerbations the features of "acute-on-chronic" inflammation have been observed. Chronic inflammation may also lead to the outlined structural changes often referred to as airway remodeling which often accounts for the irreversible component of airway obstruction observed in some patients with moderate to severe asthma and the declining lung function.

The respiratory drug market is dominated by just two indications: asthma and chronic obstructive pulmonary disease (COPD), a disease of the lower airways of the lung. National and international healthcare authorities have expressed serious concern over the rising incidences of both these diseases over the last decade. It is characterized by progressive airflow limitation which is enhanced during exacerbations. The pathological hallmarks of COPD are destruction of lung parenchyma (pulmonary emphysema) and inflammation of the peripheral (respiratory bronchiolitis) and central airways along with parenchymal inflammation of varying degree. Inflammation in COPD is associated with an inflammatory infiltrate composed of eosinophils, macrophages, neutrophils, and $CD8^+$ T lymphocytes in all lung compartments [1, 2] along with inflammatory mediators such as TNF-α, IL-8 (interleukin-8), LTB4 (leukotriene B4), ET-1 (endothelin-1) and increased expression of several adhesion molecules such as ICAM-1 [5]. The molecular mechanisms whereby inflammatory mediators are upregulated at exacerbation may be through activation of transcription factors such as nuclear factor (NF)-B and activator protein-1 that increase transcription of proinflammatory genes [6]. Acute exacerbations, linked to increased airway inflammation and oxidative stress, are the known cause of much of the morbidity, mortality, and healthcare costs associated with COPD, and they have a direct effect on disease progression by accelerating loss of lung function although the inflammatory response at exacerbation is variable and may depend in part on the etiologic agent [7, 8]. Current therapies for COPD exacerbations are of limited effectiveness [9].

Unmet Needs in COPD/Asthma Therapy

Across both asthma and COPD markets, key examples of current unmet needs include efficacious anti-inflammatory therapies for COPD (given that current therapies neither arrest nor reverse inflammation and the resulting decline in lung function), finding better ways to prevent and control asthma and COPD exacerbations, and developing therapies for the 10 % of patients with refractory asthma whose symptoms cannot be controlled with currently available drugs. In particular, there is a need to develop drugs that control the underlying inflammatory and destructive processes. Rational treatment depends on understanding the underlying disease process, and there have been recent advances in understanding the cellular and molecular mechanisms that may be involved.

Beyond the absence of curative therapy, current treatment options have inherent limitations. Despite the advances in the treatment strategies, asthma and COPD management continues to be suboptimal in many patients which is further complicated by the occurrence of exacerbations (worsening symptoms, rescue medication use, and emergency department visits or hospitalizations). Many of the orally available treatments are associated with significant adverse events. In most cases, the potential for adverse events outweighs the clinical benefit that could be derived from their long-term use as an oral agent. Thus, the successful use of oral therapies in the management of COPD has met with limited success.

Drugs most commonly used to control asthma are inhaled corticosteroids (ICSs), long-acting β_2-agonists (LABAs), methylxanthines, leukotriene modifiers, cromones, and IgE blockers. Although ICSs safely and effectively control asthma in most patients, there are numerous barriers to their use. Oropharyngeal adverse events and inadequate response to ICS in substantial number of patients present a threat to continued therapy. Hence the effectiveness of treatment with inhaled corticosteroids (ICSs) alone or combined with long-acting β_2-agonists (LABAs) comes under scrutiny. Furthermore, inhaled pharmacotherapy limits patient compliance and hence effective management of disease. Studies have shown that β_2 adrenergic agonists render receptors refractory to drug in case of repeated usage in severe chronic asthma and even combination therapy with ICS and LABA fails to reduce the exacerbations associated with recurrent episodes of asthma attack.

Targeting oxidative damage using antioxidants such as N-acetylcysteine has shown efficacy in chronic bronchitis [10] but is relatively ineffective in established COPD [11] as shown in clinical trials. Targeting TNF- to ameliorate inflammation has also been disappointing [12, 13]. The use of inhaled steroids combined with long-acting β_2 agonists to reduce exacerbation rates in more severe disease is now widely accepted, but their effects on mortality are still in doubt [13] and presently there are no effective strategies beyond smoking cessation to slow disease progression in horizon. Of concern, manipulation of the immune response shows trends to increased risk of pneumonia [14]. These data suggest that even relatively modest immunomodulators such as inhaled corticosteroids might further impact on local immunity already damaged by chronic inflammation and remodeling, rendering individuals to some degree more vulnerable to significant infections [15]. Key to effective COPD therapy is prevention of loss of alveolar smooth muscle elasticity which is irreversible by early diagnosis and more effective intervention which is currently virtually nonexistent.

In addition, the other associated debilitating airway inflammatory diseases such as airway remodeling and fibrosis are not affected by current therapy at all. Indeed, long-acting asthma therapy is efficacious, but benefits of use of asthma therapeutic drugs in COPD are inconclusive. Thus, treatment options are limited and there is a huge unmet clinical need for additional therapies [16–24]. There is an urgent need to develop an agent with bronchodilator and anti-inflammatory properties of existing anti-asthma drugs with commendable safety features. Based on this, this program aims to identify an orally efficacious agent that produces beneficial effects such as decreased disease progression and anti-inflammatory effects coupled with a good tolerability profile.

References

1. Udwadia ZF. The burden of undiagnosed airflow obstruction in India. J Assoc Phys India. 2007;55: 547–8.
2. Yang G, Rao C, Ma J, Wang L. Validation of verbal autopsy procedures for adult deaths in China. Int J Epidemiol. 2006;35(3):741–8.
3. Murphy DM, O'Byrne PM. Recent advances in the pathophysiology of asthma. Chest. 2010;137(6): 1417–26.
4. Henderson Jr WR, et al. A role for cysteinyl leukotrienes in airway remodeling in a mouse asthma model. Am J Respir Crit Care Med. 2002;165: 108–16.
5. Davidson E, Jing Jing Liub, Sheikh A. The impact of ethnicity on asthma care. Prim Care Respir J. 2010;19(3):202–8.
6. Sharm P, Halayko AJ. Emerging molecular targets for the treatment of asthma. Indian J Biochem Biophys. 2009;46(6):447–60.
7. Broide DH, Sullivan S, Gifford T, Sriramarao P. Inhibition of pulmonary eosinophilia in P selectin and ICAM deficient mouse. Am J Respir Cell Mol Biol. 1998;18(2):218–25.
8. Takizawa H. Novel strategies for the treatment of asthma. Recent Pat Inflamm Allergy Drug Discov. 2007;1:13–9.
9. Czarnobilska E, Obtułowicz K. Eosinophil in allergic and non-allergic inflammation. Przegl Lek. 2005;62(12):1484–7.
10. Erlandsen SL. Detection and spatial distribution of the beta 2 integrin (Mac-1) and L-selectin (LECAM-1) adherence receptors on human neutrophils by high-resolution field emission SEM. J Histochem Cytochem. 1993;41:327–33.
11. Poole PJ. Oral mucolytic drugs for exacerbations of chronic obstructive pulmonary disease: systematic review. BMJ. 2001;322:1271–4.
12. Decramer M. Tiotropium as a first maintenance drug in COPD: secondary analysis of the UPLIFT® trial. Lancet. 2005;365:1552–60.
13. Rennard SI. The safety and efficacy of infliximab in moderate to severe chronic obstructive pulmonary disease. Am J Respir Crit Care Med. 2007; 175(9):926–34.
14. Calverley PM. Salmeterol and fluticasone propionate and survival in chronic obstructive pulmonary disease. N Engl J Med. 2007;356:775–89.
15. Holgate ST. The epithelium takes centre stage in asthma and atopic dermatitis. TRENDS Immunol. 2007;28(6):248–51.
16. Woodside DG, Vanderslice P. Cell adhesion antagonists: therapeutic potential in asthma and chronic obstructive pulmonary disease. BioDrugs. 2008;22(2):85–100.
17. Cushley MJ. Inhaled adenosine and guanosine on airway resistance in normal and asthmatic subjects. Br J Clin Pharmacol. 1983;15:161–5.
18. Nakajima H, Sano H, Nishimura T, Yoshida S, Iwamoto I. Role of VCAM-1/VLA-4 and ICAM-1/LFA-1 interactions in antigen-induced eosinophil and T cell recruitment into the tissue. J Exp Med. 1994;179:1145–54.
19. Laberge S, Rabb H, Issekutz TB, Martin JG. Role of VLA-4 and LFA-1 in allergen –induced airway hyperresponsiveness and lung inflammation in the rat. Am J Respir Crit Care Med. 1996;151:822–9.
20. Schneider T, Issekutz TB, Issekutz AC. The role of α4 and β2 integrins in eosinophil and neutrophil migration to allergic lung inflammation in the BN rat. Am J Respir Cell Mol Biol. 1999;20:448–57.
21. Chin JE, Hatfield CA, Winterrowd GE, Brashler JR, Vonderfecht SL, Fidler SF, Griffin RL, Kolbasa KP, Krzesicki RF, Sly LM, Staite ND, Richards IM. Airway recruitment of leukocytes in mice is dependent on alpha4-integrins and vascular cell adhesion molecule-1. Am J Physiol. 1997;272:L219–29.
22. Henderson Jr WR, Chi EY, Albert RK, Chu SJ, Lamm WJ, Rochon Y, Jonas M, Christie PE, Harlan JM. Blockade of CD49d (alpha4 integrin) on intrapulmonary but not circulating leukocytes inhibits airway inflammation and hyperresponsiveness in a mouse model of asthma. J Clin Invest. 1997;100(12):3083–92. Koo GC, Shah K, et al. A small molecule VLA-4 antagonist can inhibit OVA –induced lung inflammation. Am J Respir Crit Care Med. 2003; 167:1400–9.
23. Larbi KY, Allen AR, Tam FW, Haskard DO, Lobb RR, Silva PM, Nourshargh S. VCAM-1 has a tissue-specific role in mediating interleukin-4-induced eosinophil accumulation in rat models: evidence for a dissociation between endothelial-cell VCAM-1 expression and a functional role in eosinophil migration. Blood. 2000;96:3601–9.
24. Randhawa V, Bagler G. Identification of SRC as a potent drug target for asthma, using an integrative approach of protein interactome analysis and in silico drug discovery. OMICS. 2012;16(10):513–26. Epub 2012 Jul 9.

Preclinical Models of Acute and Chronic Models of Lung Inflammation

2

Abstract

While studies have been done with various antagonists, small molecule and monoclonal and polyclonal antibodies, to inhibit a target (receptor, enzyme,peptide channel etc.) to investigate its role, saturation and specificity is always an issue. To avoid this caveat, we used a knockout mouse, so that we could validate or contradict earlier data. Of interest, often in drug discovery or target validation, a role of a molecule in a cell or a particular tissue may be investigated by chartering a certain phenomenon in its absence, that is how the physiology reestablished homeostasis despite the absence of a key molecule which ultimately defines its criticality.

Animals for Study of Role of α4 Integrin

The α4 Integrin Knockout Mouse

Gene Targeting and Generation of α4 Integrin Knockout Mice

A 13.9-kb clone containing the proximal end of the murine α4 integrin gene (α4) was isolated from a 129S4/SvJaeSor-derived genomic library (Stratagene, La Jolla, Calif.). The targeting vector was constructed from a 5.95-kb *XbaI/SphI* α4 restriction fragment that included the promoter and first two exons, a PGK-*neo*-p(A) cassette flanked by *lox*P elements, inserted at a *Kpn*I site in the promoter, and an additional *lox*P, inserted at a *Hin*dIII site distal to the second exon. The diphtheria toxin A chain gene was added as a negative selection marker (68). AK7 embryonic stem (ES) cells were maintained on mitomycin C-inactivated SNL 76/7 feeder cells in medium containing 500 U of leukemia inhibitory factor (Gibco-BRL, New York, N.Y.) per ml. A total of 8×10^6 AK7 cells were electroporated at 240 V and 500 μF with 25 μg of linearized targeting vector by using a Gene Pulser electroporator (Bio-Rad Laboratories, Hercules, Calif.); the cells were then selected in 300 μg of G418. Clones with a floxed (i.e., flanked by *lox*P) α4 allele ($α4^{flox}$) resulting from homologous recombination were identified by amplification reactions with primers specific to the distal *lox*P and a region of intron II not included in the targeting vector (5′-TGAAGAGGAGTTTACCCAGC-3′ and 5′-CACCCTTAGCTCATCATCATCG-3′). Candidate clones were analyzed by Southern blot analyses using probes proximal and distal to the 5.95-kb *XbaI/SphI* restriction fragment. Targeted clones with a normal XY karyotype were injected into C57BL/6 blastocysts 3.5 days postcoitum

E.R. Banerjee, *Perspectives in Inflammation Biology*,
DOI 10.1007/978-81-322-1578-3_2, © Springer India 2014

and transferred into pseudopregnant females. The resulting high percentage of male chimeras, identified by the level of agouti coat color, were bred to C57BL/6 females. Offspring were genotyped by tail tipping and polymerase chain reactions with primers flanking the distal *lox*P sequence (5′-GTCCACTGTTGGGCAAGTCC-3′ and 5′-AAACTTGTCTCCTCTGCCGTC-3′), which were carried out with an annealing temperature of 61 °C. Animals heterozygous for the floxed α4 allele ($\alpha 4^{flox}$) were crossed to generate floxed homozygotes.

α4 Knockout Mice and Treatment

Animal procedures were approved by the Institutional Animal Care Committee of the University of Washington. Mice (C57Bl/6.B129 background) used were Mxcre_α4f/f and Mx-cre + α4Δ/Δ (α4flox/flox bred to Mxcre + [19]). Ablation of α4 integrin in Mxcre + α4f/f mice was induced by 3 i.p. (intraperitoneal) injections of poly(I)-poly(C) (Sigma, St. Louis, MO, USA) during the first week of life. Neonatally ablated α4Δ/Δ Mxcre mice were studied as adults. β2 ko (CD18_/_) mice were previously described [10]. VCAM-1f/f mice were bred to Tie2cre + transgenics to generate VCAM-1-deficient mice as described [20]. In addition, in selected experiments, α4f/f mice also bred to Tie2cre transgenics to generate cre recombination of the α4 gene under Tie2 promoter as described [21] were used. α4 integrin$^{f/f}$ mice were produced, as previously described. These mice were bred with Mx*cre* + mice, and the resulting Mx*cre* + $\alpha 4^{flox/flox}$ mice were conditionally ablated by i.p. injections of poly(I)poly(C) (Sigma Aldrich Co., St. Louis, MO) for interferon induction. *cre*−$\alpha 4^{f/f}$ mice were used as controls and the α4-ablated mice are referred to as $\alpha 4^{\Delta/\Delta}$. All animal protocols were approved by the University of Washington (UW) IACUC. Mice were bred and maintained under specific pathogen-free conditions in UW facilities and were provided with irradiated food and autoclaved water ad libitum.

The β2 Integrin Knockout Mouse or CD18 Knockout Mouse

CD18 mice were provided by Dr. Arthur Beaudet, Baylor College of Medicine, Houston, TX

Targeting Construct and Generation of Mutant Mice

We have used a previously published (8) targeting construct with a ~ 10-kb segment of the CD18 gene containing exons 2 and 3 and a neomycin-resistance gene (neo) cassette with a short version of the RNA polymerase II promoter and the bovine growth hormone polyadenylation signal; the neomycin cassette, thus, disrupts the splice-acceptor site of exon 3 (8). The AB1

Targeting of the murine α4 *locus*. (**A**) Maps of the targeting construct containing three *lox*P elements (*gray bars*) and positive (NEO) and negative (DT-A) selection cassettes; the α4 locus ($\alpha 4^{+}$) including the first two exons (*black boxes*); the locus after homologous recombination ($\alpha 4^{flox}$) and after Cre-mediated excision ($\alpha 4^{\Delta}$). Sites for restriction enzymes *Hin*dIII (H), *Kpn*I (K), *Sph*I (S), and *Xba*I are shown; sites introduced after homologous recombination are in bold typeface. The *Xba*I restriction fragments that are diagnostic for each of the three alleles are shown. (**B**) Homologous recombination and Cre-mediated excision at the α4 locus. The $\alpha 4^{+}$ allele in ES cells migrates as a 12-kb *Sph*I (lane 1) or 7.1-kb *Xba*I (lane 3) fragment, and the $\alpha 4^{flox}$ allele migrates as a 5.5-kb *Sph*I (lane 2) or 5.7-kb *Xba*I (lane 4) fragment. BM cells from $\alpha 4^{flox/flox}$ mice were transduced with retroviruses encoding Cre and/or GFP, sorted, and cultured in methylcellulose. *Xba*I-digested DNAs from *MSCViresGFP*-transduced $\alpha 4^{+/+}$ colonies (lane 5) are compared to those from CFU-C transduced by *MSCVires-GFP* (lane 6) or *MSCV-cre-iresGFP* (lane 7). The location of the probe used is indicated by the hatched bar in panel A. (**C**) BM cells from $\alpha 4^{+/+}$, $\alpha 4^{flox/+}$, and $\alpha 4^{flox/flox}$ littermates were stained with PE-conjugated anti-α4 integrin antibody (PS/2) and analyzed by FACS (*black histograms*). An isotype control FACS profile (*dotted line*) is included for comparison. Typical FACS profiles for each group of animals are shown (*Ref.* Scott LM, Priestley GV, Papayannopoulou T. Deletion of alphα4 integrins from adult hematopoietic cells reveals roles in homeostasis, regeneration, and homing. Mol Cell Biol. 2003 Dec;23(24):9349–60)

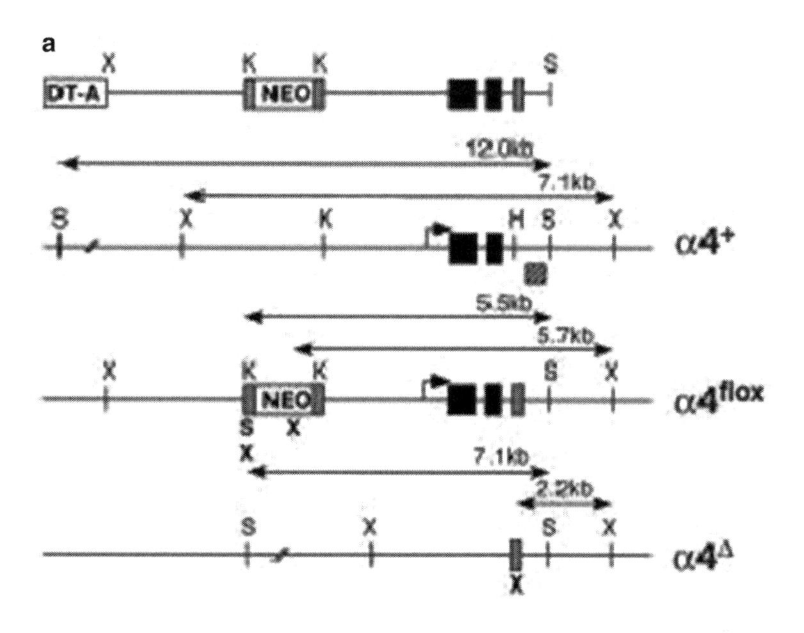
a
X
K K
S
DT-A
NEO
12.0kb
7.1kb
S X K H S X
α4+
5.5kb
5.7kb
X K K S X
NEO
S X
X
α4flox
7.1kb
2.2kb
S X S X
X
α4Δ

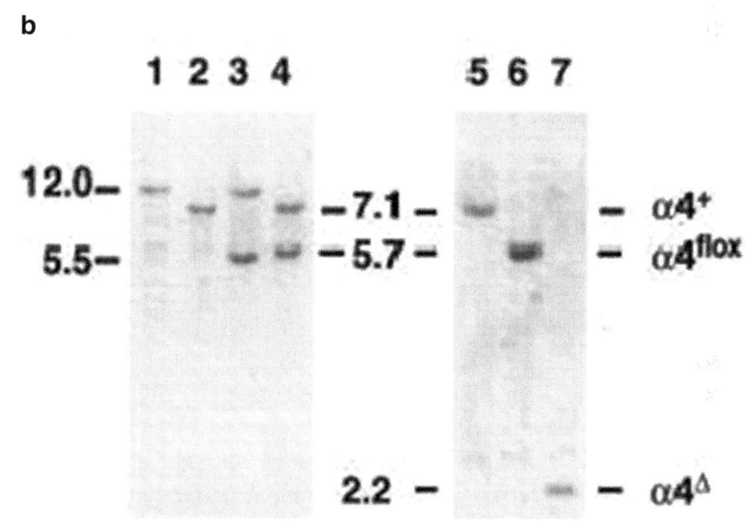
b
1 2 3 4
5 6 7
12.0
5.5
7.1
5.7
2.2
α4+
α4flox
α4Δ

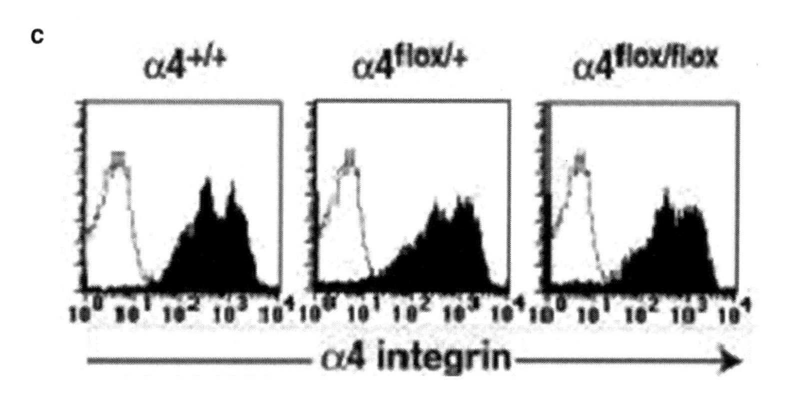
c
α4+/+
α4flox/+
α4flox/flox
α4 integrin

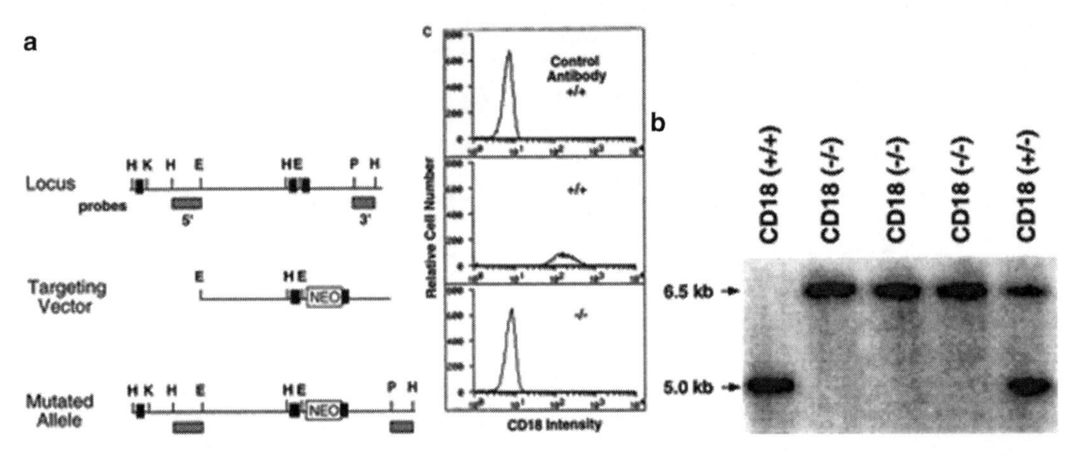

Fig. 2.1 Preparation and analysis of the CD18 null mutation. (**A**) shows the genomic region to be targeted above with the targeted mutation below. DNA length is drawn to scale and the first three exons are indicated as solid rectangles. The location of the 3′ and 5′ flanking probes is shown as hatched boxes. Restriction enzyme sites are indicated as follows: *E*, EcoRI; *H*, HindIII; *K*, KpnI; and *P*, Pst. The targeting vector was linearized with KpnI, and a replacement mutation was obtained with insertion of a neomycin cassette at a PstI site that occurs precisely at the boundary of the intron and the 5′ end of exon 3. (**B**) shows a Southern blot with the 3′ flanking probe and genomic DNA isolated from the tails of wild-type (+/+), homozygous CD18 null (−/−), and heterozygous (+/−) mice. (**C**) presents the analysis of expression of CD18 on the surface of leukocytes from wild-type and homozygous CD18 null mice. FACS® analysis was performed as described in "Materials and Methods." An isotype-matched control antibody was used (*top*). Expression of CD18 was quantitated by immunostaining with FITC-conjugated mAb C71/*16* (*Ref.* Scharffetter-Kochanek K, Lu H, Norman K, van Nood N, Munoz F, Grabbe S, McArthur M, Lorenzo I, Kaplan S, Ley K, et al. Spontaneous skin ulceration and defective T-cell function in CD18 null mice. J Exp Med. 1998; 188: pp. 119–31)

ES cell line was electroporated (13) with the construct after digestion with KpnI for use as a replacement vector. Digestion with HindIII was used to identify homologous recombinants on Southern blots hybridized with the 3′ flanking probe as indicated (Fig. 2.1a). The mutation resulted in a ~ 6.5-kb HindIII fragment compared with a 5.0-kb HindIII fragment for the wild-type genomic DNA. Embryonic stem cells confirmed by Southern blotting to carry the replacement mutation were used for germline transmission which was confirmed by Southern blot analysis of tail DNA using HindIII digestion and the 3′ flanking probe. Progeny containing the mutant CD18 allele was intercrossed resulting in CD18 homozygous mutant pups. The results presented are from mice of mixed 129/Sv and C57BL/6 J background. For all experiments, homozygous progeny and wild-type littermates from heterozygous intercrosses were used. Sentinel mice housed in the same rooms at the barrier facility as the CD18 null mice repeatedly tested negative for common viruses.

Study Design for the Development of Murine Acute Allergic Asthma Model

Female BALB/c mice (6–8 weeks of age; The Jackson Laboratory, Bar Harbor, Me) received an intraperitoneal injection of 100 mg of ovalbumin (OVA; 0.2 mL of 500 mg/mL) complexed with alum on days 0 and 14 (Fig. 2.1). Mice were anesthetized with 0.2 mL of ketamine (6.5 mg/mL)/xylazine (0.44 mg/mL) in normal saline administered intraperitoneally before receiving an intranasal dose of 50 mg of OVA (50 mL of 1 mg/mL) on days 14, 25, and 35 (Fig. 2.1). The control group received 0.2 mL of normal saline with alum administered intraperitoneally on days 0 and 14 and 0.4 mL of saline without alum administered intranasally on days 14, 25, and 35. In both the saline- and OVA-treated groups, miniosmotic pumps (200 mL, Alzet Model 2004; Durect Corp, Cupertino, Calif) containing either (R)- or (S)-albuterol (1 mg _ kg21 _ d21, 6 mL/d delivery administration) were inserted subcutaneously on the back slightly posterior to the scapulae on day 13 and remained in place until study conclusion on day 36 (Fig. 2.1). Absorption of the compounds by local capillaries results in systemic administration. Each study group consisted of 4–6 animals. The 1 mg _ kg21 _ d21 dose of albuterol enantiomer infusion was selected on the basis of prior work by Sartori et al. [9], demonstrating that continuous release of racemic albuterol (2 mg _ kg21 _ d21) subcutaneously by means of miniosmotic pump produced steady-state, high-plasma levels of albuterol (1,025 M) in mice.

Functional Validation of Preclinical Acute Allergic Asthma

In vivo airway hyperresponsiveness to methacholine was measured 24 h after the last OVA challenge by both invasive and noninvasive plethysmography.

Measuring Lung Function

Pulmonary Function Testing

1. By Unrestrained Whole-Body Plethysmography

   In vivo airway responsiveness to methacholine was determined on day 36 in conscious, freely moving, spontaneously breathing mice by using whole-body plethysmography (Model PLY 3211; Buxco Electronics Inc, Sharon, Conn), as described by Hamelmann et al. [10]. Mice were challenged with aerosolized saline or increasing doses (2 and 10 mg/mL) of methacholine generated by an ultrasonic nebulizer (DeVilbiss Health Care, Inc, Somerset, Pa) for 2 min.

   The degree of bronchoconstriction was expressed as enhanced pause (Penh), a calculated dimensionless value that correlates with measurement of airway resistance, impedance, and intrapleural pressure [9–11]. Penh readings were taken and averaged for 4 min after each nebulization challenge. Penh is calculated as follows: Penh ¼ ½ðTe = Tr21 + 3ðPEF = PIF + _, where Te is expiration time, Tr is relaxation time, PEF is peak expiratory flow, and PIF is peak inspiratory flow 3 0.67 coefficient. The time for the box pressure to change from a maximum to a user-defined percentage of the maximum represents the relaxation time. The Tr measurement begins at the maximum box pressure and ends at 40 %. Because Penh is the ratio of measurements obtained during the same breath, it is mainly independent of functional residual capacity, tidal volume, and respiratory rate.

2. Invasive Plethysmography

   On day 22, 24 h after the last intratracheal allergen (OVA) challenge, invasive pulmonary mechanics were measured in mice in response to methacholine in the same manner as previously described [13] with the following modifications: (a) the thorax was not opened, (b) mice were ventilated with a tidal volume of 20 μl and respiratory rate of 120 breaths/min using a MiniVent Ventilator for Mice (Harvard Apparatus, Holliston, MA), (c) mice received aerosolized solutions of methacholine (0, 3.125, 6.25, 12.5, 25, 50, and 100 mg/ml in normal saline) via an AER 1021 nebulizer aerosol system (Buxco Electronics, Inc., Wilmington, NC) with 2.5–4 μm aerosol particle size generated by NEB0126 nebulizer head (Nektar Therapeutics, San Carlos, CA), and (d) a commercial plethysmography system (Model PLY4111 plethysmograph, MAX II amplifier and pressure transducer system, and Biosystem XA software, Buxco Electronics, Inc.) was used to determine RL as calculated from measures of pressure and flow and expressed as cmH_2O/ml/s). Noninvasive plethysmography (expressed as Penh) was also assessed on day 22 in independent experiments.

Measuring Other Functional Parameters by Morphometry

Histological Assessment of Airway and Lung Parenchyma Damage

After pulmonary function testing, bronchoalveolar lavage (BAL) was performed on the right lung, with total BAL fluid cells counted and eosinophils identified by means of eosin staining [12]. Left lung tissue was obtained for histopathology, and plasma was obtained for

OVA-specific IgE levels. Ten lung sections per animal were randomly selected and examined in a blinded manner. Sections were stained with hematoxylin and eosin, the total inflammatory cell infiltrate was assessed on a semiquantitative scale (0–41), the number of eosinophils per unit of airway area (2,200 mm^2) was determined by using a point-counting system (Image-Pro Plus point-counting system software, Version 1.2 for Windows; Media Cybernetics, Silver Spring, Md) [12], and interstitial and perivascular airway edemas were assessed [13,14]. Airway goblet cells (as a percentage of total airway cells) were identified by means of Alcian blue staining [12], and the degree of mucus plugging of the airways (0.5–0.8 mm in diameter) with the percentage occlusion of the airway diameter was classified on a 0 to 41 scale on the basis of the following criteria: 0, no mucus; 1, approximately 10 % occlusion; 2, approximately 30 % occlusion; 3, approximately 50 % occlusion; and 4, greater than approximately 80 % occlusion [12]. Cytokine assays IL-4, IL-5, IL-10, GM-CSF, TNF-α, IL-2, and IFN-γ were assayed in BAL fluid with Bio-Plex Mouse Cytokine assays (Bio-Rad Laboratories, Hercules, Calif) that are bead-based multiplex sandwich immunoassays with a limit of detection of less than 10 pg/mL. IL-13 was assayed in BAL fluid with a mouse IL-13 immunoassay (Quantikine M; R&D Systems, Minneapolis, Minn), with a limit of detection of less than 1.5 pg/mL. OVA-specific IgE was assayed by modification of the method of Iio et al. [15]. Nunc 96-well flat-bottom plates (Nalge Nunc International, Rochester, NY) were coated with 50 mg/mL OVA in 13 PBS overnight at room temperature, washed 3 times with 13 PBS plus 0.05 % Tween-20 (wash buffer), blocked with 3 % BSA in 13 PBS for 1 h at room temperature, and washed 4 times with wash buffer. Fifty-microliter plasma samples (1:1 in 13 PBS) were added per well and incubated for 90 min at 37_C, then washed 4 times with wash buffer, and blotted dry by inverting over paper towels. One hundred microliters (1:100 in 13 PBS) of biotin-conjugated rat anti-mouse IgE mAb (clone R35-72; BD Biosciences, San Diego, Calif) was added to each well and incubated overnight at 4_C and then washed 4 times with wash buffer and blotted dry. Then 100 mL per well (1:1000 in 13 PBS) streptavidin-horseradish peroxidase-conjugated secondary antibody (BD Biosciences) was added, and samples were incubated at 37_C for 90 min, then washed 4 times with wash buffer, and blotted dry. One hundred microliters of substrate solution (i.e., 1 tablet of 2,2′-azino-bis [3-ethylbenzthiazoline-sulfonic acid], ABTS; Sigma Chemical Co, St Louis, Mo] dissolved in 100 mL of 0.05 M phosphate-citrate buffer, pH 5.0, and 25 mL of 30 % H_2O_2) was added per well, and color was developed for 30 min at room temperature. OD 405 nm was measured by using an AD 340C absorbance detector (Beckman Coulter Inc, Fullerton, Calif). For the IgE standard curve, a sandwich ELISA was used in which in separate assay plates biotin-conjugated rat anti-mouse IgE mAb (clone R35-72, BD Biosciences) was used to coat the wells, and instead of plasma samples, known concentrations of purified anti-mouse IgE (clone C38-2, BD Biosciences) were incubated; assays were run as described above. The standard curve was constructed by using a linear regression analysis of the absorbances against serial dilutions of known concentrations of mouse IgE. Pooled mouse plasma from OVA-sensitized/OVA-challenged mice was used as a positive control. In "Statistical Analysis," the data are reported as the means ± SE of the combined experiments. Differences were analyzed for significance ($P < .05$) by means of ANOVA with the protected least-significant-difference method (StatView II; Abacus Concepts, Berkeley, Calif).

Lung Histology: Histochemistry

Lungs from other animals of the same group were fixed in 4 % paraformaldehyde overnight at 4 °C. The tissues were embedded in paraffin and cut into 5-μm sections. A minimum of 15 fields were examined by light microscopy. The intensity of cellular infiltration around pulmonary blood vessels was assessed by hematoxylin and eosin (H&E) staining. Airway mucus was identified by staining with Alcian blue and periodic acid Schiff staining as described previously (3).

Subepithelial pulmonary fibrosis was detected by Masson's trichrome and Martius scarlet blue stains as described in (10).

Lung Immunohistochemical Staining

Lungs were processed for immunohistochemical staining following standard procedure (18) and then stained with either anti-VCAM-1 (vascular cell adhesion molecule-1, MK2), anti-β1 (9EG7), or anti-TGF-β1. Briefly, tissues were fixed with 4 % paraformaldehyde in 100-mM PBS (phosphate buffered saline, pH 7.4) for 6–12 h at 4 °C; washed with PBS for 10 min 3 times and then soaked in 10 % sucrose in PBS for 2–3 h, 15 % sucrose in PBS for 2–3 h, and 20 % for 3–12 h at 4 °C; and then embedded in OCT compound (Tissue-Tek 4583, Sakura Finetechnical Co., Ltd, Tokyo, Japan) and frozen in acetone-cooled dry ice. Frozen blocks were cut on a freezing, sliding microtome at 4 μm (Leica CM1850 Cryostat) and air-dried for 30 min at RT. After washing in PBS 3 times for 10 min at RT, 0.3 % hydrogen peroxide was applied to each section for 30 min at RT to block endogenous peroxidase activity. Each slide was incubated with blocking solution to block nonspecific reactions, and appropriately diluted primary antibody was applied to each slide and incubated overnight at 4 °C. After washing with PBS, slides were incubated with appropriately diluted specific biotin-conjugated secondary antibody solution for 1 h at RT. After washing with PBS, slides were incubated in AB reagent for 1 h at RT (ABComplex/HRP, DAKO, Carpinteria, CA), washed with PBS, and stained with 0.05 % DAB (3,3′-diaminobenzidine tetrahydrochloride, Sigma Aldrich Co.) in 0.05 M Tris buffer (pH 7.6) containing 0.01 % H_2O_2 for 5–40 min at RT. Slides were counterstained with Mayer's hematoxylin, dehydrated, and mounted.

Cellular Traffic by Light Microscopy

(a) Bronchoalveolar Lavage Fluid (BALf)
Mice underwent exsanguination by intraorbital arterial bleeding and then lavaged (0.4 ml three times) from both lungs. Total BAL fluid cells were counted from a 50-μl aliquot, and the remaining fluid was centrifuged at 200 × g for 10 min at 4 °C, and the supernatants were stored at −70 °C for assay of BAL cytokines later. The cell pellets were resuspended in fetal bovine serum (FBS), and smears were made on glass slides. The cells, after air-drying, were stained with Wright-Giemsa (Biochemical Sciences, Inc., Swedesboro, NJ) and differential counts enumerated using a light microscope at 40X magnification. Cell number refers to that obtained from lavage of both lungs/mouse.

(b) Lung Parenchyma (LP) Cell Recovery
Lung mincing and digestion was performed after lavage as described previously (33) with 100 μg/ml collagenase for 1 h at 37 °C and filtered through a 60# sieve (Sigma Aldrich Co.). All numbers mentioned in this chapter refer to cells obtained from one lung/mouse.

(c) Fluorescein-Activated Cell Sorter (FACS) Analysis of Cells from Various Tissues
Assessing cell traffic from their site of poiesis to their homing sites or foci of inflammation in tissues, cells were isolated following various protocols from different tissues tracing the entire route the inflammatory cells traverse. Thus cells from hemolyzed peripheral blood (PB), bone marrow (BM), bronchoalveolar lavage (BAL), lung parenchyma (LP), spleen, mesenteric lymph nodes (MLN), cervical lymph nodes (CLN), axillary lymph nodes (LNX), and inguinal lymph nodes (LNI) were analyzed on a FACSCalibur (BD Immunocytometry Systems, San Jose, CA) using the CEllQuest program. Staining was performed by using antibodies conjugated to fluorescein isothiocyanate (FITC), phycoerythrin (PE), allophycocyanin (APC), peridinin-chlorophyll-protein (Per CP-Cy5.5), and Cy-chrome (PE-Cy5 and PE-Cy7). The following antibodies (BD Biosciences-Pharmingen, San Diego, CA) were used for cell-surface staining: APC-conjugated CD45 (30 F-11), FITC-conjugated CD3(145-2C11), PE-Cy5-conjugated CD4 (RM4-5), PE-conjugated CD45RC (DNL-1.9), APC-conjugated CD8(53-6.7), PE-Cy5-conjugated β220 (RA3-6β2), FITC-conjugated IgM,

PE-conjugated CD19 (ID3), PE-conjugated CD21(7G6), FITC-conjugated CD23 (B3B4), APC-conjugated GR-1(RB6-8C5), and PE-conjugated Mac1(M1/70). PE-Cy5-conjugated F4/80 (Cl:A3-1[F4/80]) was obtained from Serotec Ltd., Raleigh, NC. PE-conjugated anti-α4 integrin (PS2) and anti-VCAM-1(M/K-2) were from Southern Biotechnology, Birmingham, AL. Irrelevant isotype-matched antibodies were used as controls. Among hematopoietic cells CD45+/CD3+ were T cells, CD3+/CD4+ were helper T cells, and CD3+/CD8+ were cytotoxic T cells. B cells were β220+. Gr-1+/F4/80− cells were granular cells (neutrophils, eosinophils) and Gr-1−/F4/80^{hi} cells were tissue macrophages.

Functional Parameters of the TH1 Response

(a) Cytokines
Cytokines (IL-2, IL-4, IL-5, TNFα, and IFNγ) in BALf and serum were assayed by FACS with Mouse Th1/Th2 Cytokine Cytometric Bead Assay (BD Biosciences) following the manufacturer's protocol. The manufacturer's sensitivity for IL-2, IL-4, and IL-5 is 5 pg/mL, for IFNγ is 2.5 pg/mL, and for TNF-α is 6.3 pg/mL. IL-13 and eotaxin were measured by ELISA using Quantikine M kits (R&D Systems, Minneapolis, MN), and the limit of detection is 1.5 pg/mL for IL-13 and 3 pg/mL for eotaxin.

(b) ELISPOT Assay of Th2-Secreting Cells
IL-4-positive and IFN-γ-positive cells in single-cell suspensions from lung parenchyma and BALf were detected employing standard ELISPOT assays [10] using detection and capture monoclonal antibodies and AEC substrate reagent from BD Biosciences. Dots were counted manually using 40X magnification. Measurement of cytokines Th1/Th2 cytokines in BALf and plasma were assayed with mouse cytometric bead array (CBA, BD Biosciences). IL-13 and eotaxin were measured by Quantikine M kits from R&D Systems (Minneapolis, MN, USA). OVA-specific IgE and IgG1 in plasma anti-mouse IgE (R35-72) and IgG1 (A85-1) antibodies (BD Biosciences) were used to measure OVA-specific IgE and IgG1 in plasma using sandwich ELISA [26,27].

(c) OVA-Specific IgE and IgG1 in Plasma
Anti-mouse IgE (R35-72) and IgG1 (A85-1) from BD Biosciences were used for measuring OVA-specific IgE and IgG1, respectively, by standard ELISA procedures as previously described (16). The lower and upper limits of detection for IgE and IgG1 are 3 ng/ml to 10 ug/ml (minimal detectable dose determined by adding two standard deviations of the mean OD 405 nm value for 20 replicates of the zero standard and calculating the corresponding concentration).

(d) Soluble VCAM-1 and Soluble Collagen in Lung Homogenate
Soluble VCAM-1 was determined, as previously described (3). The total amount of soluble collagen in the lung was measured using a Sircol collagen assay kit from Biocolor, (Newtownsbury, Northern Ireland, UK), according to the method described (18). In all experiments, a collagen standard was used to calibrate the assay.

(e) Adoptive Transfer Experiment to Assess the Role of a Particular Cell Subset
At day 8 after i.p. sensitization, 5 _ 106 CD4+ splenocytes from both α4+/+ controls (α4f/fcre_) or α4Δ/Δ mice were purified by magnetic-activated cell sorting (MACS, Miltenyi Biotec, Auburn, CA, USA) and then injected into the tail veins of naïve controls or α4Δ/Δ recipients. The mice were subsequently challenged with 3 i.t. instillations of OVA over the next 72 h and sacrificed 24 h after the last instillation [28]. One group of α4Δ/Δ mice also received unpurified α4+/+ splenocytes from sensitized mice.

(f) CFU-C Assay
To quantitate committed hematopoietic progenitors of all lineages, CFU-C assays were performed as described [19]. Proliferation assay MACS-separated CD4+ and CD8+ T cells from spleens were stimulated

in vitro with various concentrations of stimuli (CD3/CD28, phorbol myristic acetate (PMA)/ionomycin, irradiated antigen-presenting cells (APCs), and lipopolysaccharide (LPS)) to assay proliferative responses. After 72 h, proliferation was measured either by CellTiter96 assay from Promega (Madison, WI, USA) measuring OD at 570 nm or by [3H]-thymidine incorporation following standard protocols.

Study Design for the Development of Murine Chronic Allergic Asthma Model

Induction of Chronic Allergic Asthma

Mice were sensitized and later challenged with OVA (Pierce Biotechnology, Inc., Rockford, IL) as described in an earlier publication (22). Briefly, mice were immunized with 100 μg OVA complexed with aluminum sulfate (Alum) in a 0.2-ml volume, administered by i.p. injection on day 0. On day 8 (250 μg of OVA) and on days 15, 18, and 21 (125 μg of OVA), mice anesthetized briefly with inhalation of isoflurane in a standard anesthesia chamber were given OVA by i.t. administration. Intratracheal challenges were done as described previously (18). Mice were anesthetized and placed in a supine position on a board. The animal's tongue was extended with lined forceps and 50 μl of OVA (in the required concentration) was placed at the back of its tongue. The control group [$\alpha 4^{+/+}$*cre*− mice also injected with poly(I)poly(C)] received normal saline with aluminum sulfate by i.p. route on day 0 and 0.05 ml of 0.9 % saline by i.t. route on days 8, 15, 18, and 21. We have previously shown that this protocol results in increased AHR, inflammation of the airways, and Th2 cytokine production (18, 19, 22). Prolonged inflammation was induced by subsequent exposure of mice to 125 μg of OVA three times a week until groups of mice were killed on day 55 (chronic phase) after the last i.t. challenge on day 54. Control groups were only given saline alum on all the treatment groups in parallel.

Mouse Model of Bleomycin-Induced Pulmonary Fibrosis

A single intratracheal dose of 0.075 U/ml of bleomycin in 40-μl saline was administered (day 0), and mice were sacrificed 14 and 21 days later. Mice C57Bl/6 were kept under ABSL-2 conditions approved by the IACUC of the University of Washington and monitored daily. They were housed under specific pathogen-free condition and were given food and water ad libitum. They were sacrificed on day 14. One week after bleomycin administration, mice developed marked interstitial and alveolar fibrosis, detected in lung sections by Masson's trichrome stain. Analysis of cell populations by enzymatic digestion by collagenase type IV followed by cell counting in Coulter counter and subsets identified and quantified by FCM and total and differential count of H&E stained cytospin smears of single-cell suspensions show loss of type II and type I alveolar epithelial cells and influx of macrophages. AEI and II were isolated following standard protocol.

Lung Function Testing

In vivo airway hyperresponsiveness to methacholine was measured 24 h after the last OVA challenge in conscious, free-moving, spontaneously breathing mice using whole-body plethysmography (model PLY 3211, Buxco Electronics, Sharon, CT) as previously described (22). Mice were challenged with aerosolized saline or increasing doses of methacholine (5, 20, and 40 mg/ml) generated by an ultrasonic nebulizer (DeVilbiss Health Care, Somerset, PA) for 2 min. The degree of bronchoconstriction was expressed as enhanced pause, a calculated dimensionless value, which correlates with the measurement of airway resistance, impedance, and intrapleural pressure in the same mouse. P_{enh} readings were taken and averaged for 4 min after each nebulization challenge. P_{enh} was calculated as follows: $P_{enh} = [(Te/Tr\text{-}1) \times (PEF/PIF)]$, where T_e is expiration time, T_r is relaxation time, PEF is peak expiratory flow, and PIF is peak inspiratory flow × 0.67 coefficient. The time for the box pressure to change from a maximum to a user-defined percentage of the maximum represents the relaxation time. The T_r measurement begins at the maximum box pressure and ends at 40 %.

Preparation of Cell Suspensions

Bone marrow cells were flushed from femurs [24]. Mice were bled retro-orbitally and red cells hemolyzed as described [24]. Bronchoalveolar lavage fluid (BALf) was collected as described [23]. Lung parenchymal cells (LP) were collected from lavaged lungs that were minced and digested as described in [13] and then vortexed and filtered through sterile 40-mm Nitex filter (Sefar America, Depew, NY, USA) before use.

BALf

After pulmonary function testing, the mouse underwent exsanguination by intraorbital arterial bleeding and then BAL (0.4 ml three times) of both lungs. Total BAL fluid cells were counted from a 50-μl aliquot and the remaining fluid was centrifuged at 200 *g* for 10 min at 4 °C, and the supernatants were stored at −70 °C for assay of BAL cytokines later. The cell pellets were resuspended in FCS and smears were made on glass slides. The cells, after air-drying, were stained with Wright-Giemsa (Biochemical Sciences Inc, Swedesboro, NJ), and their differential count was taken under a light microscope at 40 × magnification. Cell number refers to that obtained from lavage of both lungs/mouse.

Lung Parenchyma

Lung mincing and digestion was performed after lavage as described previously [11] with 100 u/ml collagenase for 1 h at 37 °C and filtered through a #60 sieve (Sigma). All numbers mentioned in this chapter refer to cells obtained from one lung/mouse. The cells recovered are primarily from the parenchyma as the lungs prior to lavage and then enzymatic separation of parenchyma cells were thoroughly exsanguinated after ligating the vena cava to stop drainage into pulmonary circulation.

Lung Histology

Lungs of other animals of same group were fixed in 4 % paraformaldehyde overnight at 4 °C. The tissues were embedded in paraffin and cut into 5-μm sections. A minimum of 15 fields were examined by light microscopy. The intensity of cellular infiltration around pulmonary blood vessels was assessed by hematoxylin and eosin staining. Airway mucus was identified by staining with Alcian blue and periodic acid Schiff staining as described previously [12].

Goblet Cell Metaplasia

Using a microscope, percentages of metaplastic goblet cells (detected by Alcian blue staining) in lung sections were calculated with counts from 10 different fields per mouse as described [22].

Fluorescein-Activated Cell Sorter (FACS) Analysis

Cells from hemolyzed peripheral blood (PB), bone marrow(BM), bronchoalveolar lavage (BAL), lung parenchyma (LP), spleen, mesenteric lymph nodes (MLN), cervical lymph nodes (CLN), axillary lymph nodes (LNX), and inguinal lymph nodes (LNI) were analyzed on a FACSCalibur (BD Immunocytometry Systems, San Jose, CA) by using the CEllQuest program. Staining was performed by using antibodies conjugated to fluorescein isothiocyanate (FITC), phycoerythrin (PE), allophycocyanin (APC), peridinin-chlorophyll-protein (Per CP-Cy5.5), and Cy-chrome (PE-Cy5 and PE-Cy7). The following BD Pharmingen (San Diego, CA) antibodies were used for cell-surface staining: APC-conjugated CD45 (30 F-11), FITC-conjugated CD3(145-2C11), PE-Cy5-conjugated CD4 (RM4-5), PE-conjugated CD45RC (DNL-1.9), APC-conjugated CD8(53-6.7), PE-Cy5-conjugated β220 (RA3-6β2), FITC-conjugated IgM, PE-conjugated CD19 (ID3), PE-conjugated CD21(7G6), FITC-conjugated CD23 (B3B4), APC-conjugated GR-1(RB6-8C5), and PE-conjugated Mac1(M1/70). PE-Cy5-conjugated F4/80 (Cl:A3-1(F4/80)) was obtained from Serotec Ltd., Oxford, UK. PE-conjugated anti-α4 integrin (PS2) and anti-VCAM-1(M/K-2) were from Southern Biotechnology, Birmingham, Ala. Irrelevant isotype-matched antibodies were used as controls. As for the nonhematopoietic cells of the lung, the following markers were used from Santa Cruz, CA, USA: SP-C, Oct-3/4, SSEA-3 & 4, TTF-1, AQP-1, and AQP-5 as unlabelled primary antibodies, and a FITC-labeled secondary antibody was used for FACs detection. So FACS surface marker expression experiments are all single-marker analysis not combinatorial. 10^6 cells were taken per sample in 50 ml cell suspension in ice cold PBS (16); 10^5 events were recorded per sort [13]. To eliminate differences in expression of activation markers by the choice of enzymes used for isolation, a comparative was done with collagenase, trypsin, and dispase was found to be least modulating of marker expression than the other two digestion by which resulted in higher BrdU cells and less SP showing that somehow cells were effluxing dyes at a higher rate when digested thus. At this stage we can only speculate as to the effect of these enzymes on marker expression [14].

Proliferation Assay

MACS-separated CD4+ and CD8+ T cells from spleens were stimulated in vitro with various concentrations of stimuli (CD3/CD28, phorbol myristic acetate (PMA)/ionomycin, irradiated antigen-presenting cells (APCs), and lipopolysaccharide (LPS) to assay proliferative responses. After 72 h, proliferation was measured either by CellTiter96 assay from Promega (Madison, WI, USA) measuring OD at 570 nm or by [3H]-thymidine incorporation following standard protocols.

CFU-C Assay

To quantitate committed progenitors of all lineages, CFU-C assays were performed using methylcellulose semisolid media (Stemgenix, Amherst, N.Y.) supplemented with an additional 50 ng of stem cell factor (Peprotech, Rocky Hill, N.J.) per ml. Next, 50,000 cells from bone marrow, 500,000 cells from spleen, 0.01 million cells from lung and BAL, and 10 μl peripheral blood were plated on duplicate 35-mm culture dishes and then incubated at 37 °C in a 5 % CO_2-95 % air mixture in a humidified chamber for 7 days. Colonies generated by that time were counted by using a dissecting microscope, and all colony types (i.e., burst-forming units-erythroid [BFU-e], CFU-granulocyte-macrophage [CFU-GM], and CFU-mixed [CFU-GEMM]) were pooled and reported as total CFU-C. Total CFU-C per organ was calculated by extrapolating CFU-C against number of plated cells to the total number of cells in the organ.

Study Design to Identify Resident Stem Cells of the Lung

BrdU Pulse Chase

BrdU is a DNA analogue. Slow-cycling cells are assumed to be stem cells and pulsing of control versus bleomycin (single i.t. dose of 0.075 U/ml bleomycin) treated WT C57Bl/6 mice over 2, 4, and 6 days i.p. at 12-h interval, and chase over 10 weeks is expected to yield BrdU positive cells and negative cells. While negative cells are assumed to be mature regularly cycling cells, BrdU + cells after 10 weeks of chase are most likely slow-cycling stem cells that started cycling late and hence retain the label latest (label-retaining cells or LRC). C57Bl/6 mice were intratracheally instilled with 0.075U/ml bleomycin in 40-μl volume under brief isoflurane anesthesia, and animals were maintained under SPF conditions in the UW animal facilities and sacrificed periodically to assess the above. The abbreviations used are the following: i.p. intraperitoneal; i.t. intratracheal; BAL, bronchoalveolar lavage; and PB, peripheral blood.

Isolation of SP Cells

Lung parenchyma and cells migrated to airways were isolated following collagenase digestion of 1-mm pieces of lung tissue. These were incubated with Hoechst dye for 30 min at 37 °C and then after washing were kept for 90 min at RT followed by sorting of "side population cells" which are the late effluxing putative stem cells of the lung.

Statistics

Mean ± S.E.M. (standard error of the mean) was calculated using Student's *t*-test in Excel Software (Microsoft, Redmond, WA). $p < 0.05$ was considered statistically significant.

Studying the Roles of Some Key Molecules in Acute Allergic Asthma

3

Abstract

Patients are usually prescribed a racemic mixture of S and R albuterol. The aim of the study was to delineate functional efficacy of either or both of them in acute allergic asthma using a murine model. While both R and S albuterol reduce airway eosinophil trafficing and mucus hypersecretion in the mousemodel of asthma, S-albuterol increases allergen-induced airway edema and hyperresponsivness which limits the clinical efficacy of the S enantiomerin over-the-counter drugs for asthma.

Research Area 1: Enantiomers of Albuterol, the OTC Drug of Choice for Acute Asthma Management

Investigating the Therapeutic Potential and Functional Index of Two Enantiomers of Albuterol, the Drug of Choice in Clinical Asthma in a Mouse Allergic Asthma Model

Background and Relevance of the Study

Adrenergic receptors are composed of α- and β-receptors that bind endogenous catecholamines, such as epinephrine. Although three subtypes of b-adrenergic receptors exist, smooth muscle relaxation producing vasodilation and bronchodilation is mediated by the β2-receptor. Short-acting β2-adrenergic receptor agonists rapidly induce bronchodilation in patients with asthma and are used for relief of acute symptoms, prevention of exercise-induced asthma, and management of acute severe asthma. Racemic albuterol contains equal concentrations (50:50) of the (R)- and (S)-enantiomers (i.e., enantiomers that are nonsuperimposable mirror images) [1]. The (R)-enantiomer of albuterol binds to β2-adrenergic receptors with nearly 100-fold greater affinity than the (S)-enantiomer, suggesting that the (S)-enantiomer does not act through b-adrenergic receptor activation [1]. Whereas the (R)-enantiomer of albuterol (levalbuterol) exerts the bronchodilating properties of albuterol, the (S)-enantiomer has adverse effects, including augmentation of bronchospasm and proinflammatory activities [2–5]. In murine mast cells (S)-albuterol increases IgE-induced histamine and IL-4 production, whereas the (R)-enantiomer lacks these effects [2]. Anti-inflammatory effects of (R)-albuterol, such as inhibition of T-cell proliferation, might be negated by the presence of the (S)-enantiomer [3]. Differential effects of the enantiomers might result from differences in pharmacokinetics [1]. The initial step in the metabolism of the (S)-

E.R. Banerjee, *Perspectives in Inflammation Biology*,
DOI 10.1007/978-81-322-1578-3_3, © Springer India 2014

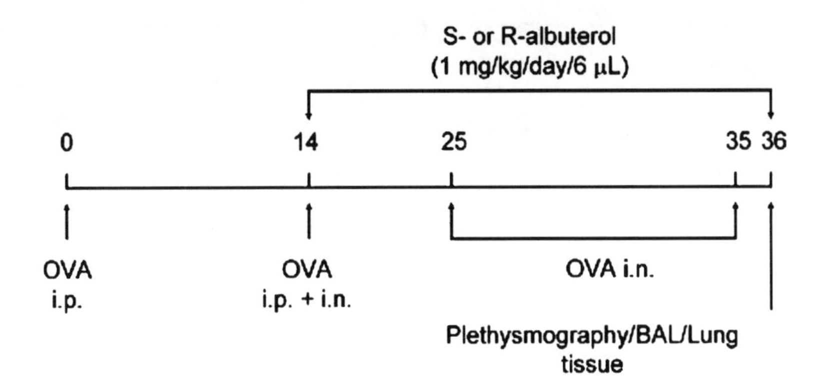

Fig. 3.1 Study protocol. *i.p.* intraperitoneal, *i.n.* intranasal

and (R)-enantiomers is sulfate conjugation, a stereospecific process in human airway epithelial cells and other cells and tissues [6]. The greater rate of sulfate conjugation of (R)-albuterol might lead to lower plasma levels of (R)- than (S)-albuterol in human subjects [7]. Potential adverse effects of (S)-albuterol on asthma control might also be augmented by increased binding to lung tissue. In this study, we characterized the effects of the (R)- and (S)-enantiomers of albuterol on allergic airway inflammation and hyperresponsiveness in a mouse asthma model that mimics key features of human asthma [8]. Although prior studies in guinea pigs and human subjects have demonstrated that the (S)-enantiomer of albuterol can induce airway hyperreactivity, there are no prior studies examining the effect of (S)-albuterol versus (R)-albuterol on both airway hyperresponsiveness and the TH2 phenotype (i.e., allergen-induced airway eosinophil trafficking, mucus metaplasia, edema, and TH2 cytokine release) in an in vivo asthma model. We report that both (R)- and (S)-enantiomers reduce allergen-induced airway eosinophil and mucus gland hyperplasia. However, only (S)-albuterol increases airway edema and responsiveness to methacholine, effects that would decrease the therapeutic efficacy of racemic albuterol.

Results in a Nutshell

In OVA-sensitized/OVA-challenged mice, (R)-albuterol significantly reduced the influx of eosinophils into the bronchoalveolar lavage fluid and airway tissue. (R)-Albuterol also significantly decreased airway goblet cell hyperplasia and mucus occlusion and levels of IL-4 in bronchoalveolar lavage fluid and OVA-specific IgE in plasma. Although (S)-albuterol significantly reduced airway eosinophil infiltration, goblet cell hyperplasia, and mucus occlusion, it increased airway edema and responsiveness to methacholine in OVA-sensitized/OVA-challenged mice. Allergen-induced airway edema and pulmonary mechanics were unaffected by (R)-albuterol (Fig. 3.1).

Detailed Results

Effect of (R)- and (S)-Enantiomers of Albuterol on Allergen-Induced Airway Inflammation Airway Infiltration by Eosinophils

A marked infiltration of inflammatory cells that were predominantly eosinophils around the airways and pulmonary blood vessels was observed in the lung interstitium of OVA-treated mice (Fig. 3.2b) compared with that seen in saline-treated control mice (Fig. 3.2a) on day 36, 24 h after the last intranasal OVA or saline challenge. By means of morphometric analysis (Fig. 3.3a, b) of the histological sections (Fig. 3.2c and d vs. b), administration of (R)- and (S)-albuterol by means of miniosmotic pumps (1 mg _ kg21 _ d21 dose from days 13 to 36) significantly decreased the influx of total inflammatory cells ($P = 0.0035$, R-albuterol/OVA vs. OVA; $P = 0.0226$, S-albuterol/OVA vs. OVA; Fig. 3.3a) and eosinophils ($P = 0.021$, R-albuterol/OVA vs. OVA; $P = 0.008$; S-albuterol/OVA vs. OVA; Fig. 3.3b) into the lung interstitium.

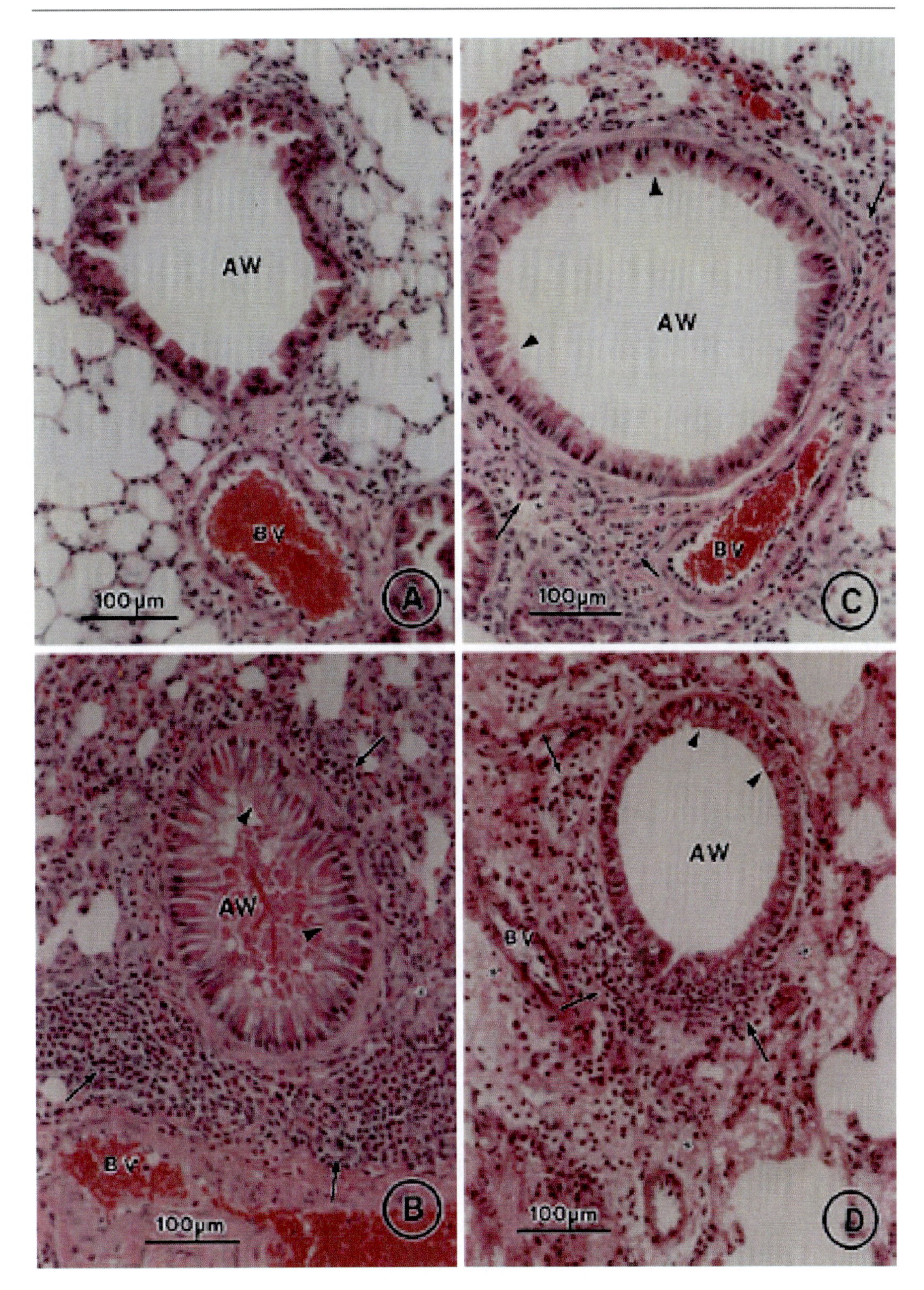

Fig. 3.2 Effect of (R)- and (S)-albuterol enantiomers on airway histopathology in a mouse asthma model. Lung tissue was obtained on day 36 from saline-treated control animals (**a**), OVA-treated control animals (**b**), OVA-treated mice administered (**R**)-albuterol (**c**), and OVA-treated mice administered (**S**)-albuterol (**d**), and sections were stained with hematoxylin and eosin. *Arrows* indicate eosinophils and other inflammatory cells, *arrowheads* indicate mucus, and *asterisks* indicate edema. *AW* airway, *BV* blood vessel. *Bars* = 100 µm

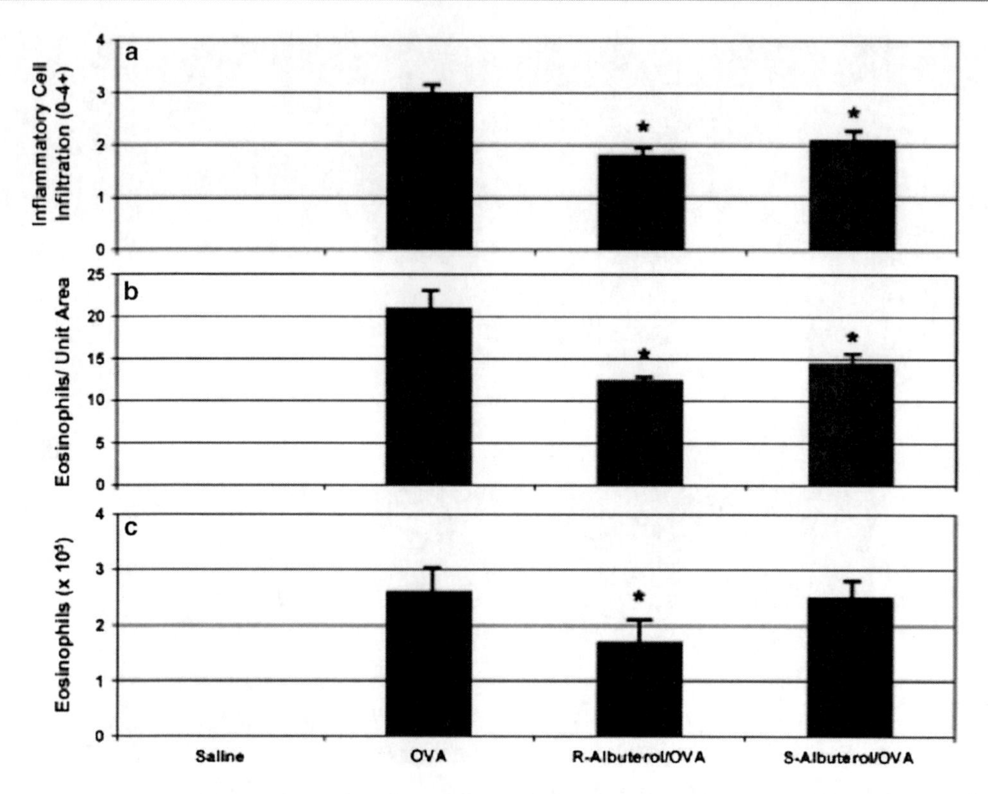

Fig. 3.3 Effect of (R)- and (S)-albuterol enantiomers on allergen-induced airway inflammatory cell infiltration. The total inflammatory cell infiltration of the airways (**a**), the number of eosinophils per unit area (2,200 μm^2) of lung tissue (**b**), and eosinophils per milliliter of BAL fluid (**c**) were determined. $^*P < 0.05$ versus OVA

Compared with the saline group (Fig. 3.3c), OVA-sensitized/OVA-challenged mice exhibited a marked increase in BAL fluid eosinophils to 2.5 6 0.5 3 105 eosinophils/mL ($P < 0.0001$, OVA vs. saline; Fig. 3.3c), which represented 41.8 % of total BAL fluid cells. In OVA-sensitized/OVA-challenged mice, treatment with (R)-albuterol significantly inhibited the influx of eosinophils into BAL fluid by 40.6 % ($P = 0.0043$, R-albuterol/OVA vs. OVA; Fig. 3.3c). In contrast, (S)-albuterol had no significant effect on eosinophil influx into the BAL fluid of OVA-treated mice (Fig. 3.3c).

Airway Mucus Hypersecretion. Hyperplasia of airway goblet cells and hypersecretion of mucus were observed in OVA-treated mice (Fig. 3.2b) compared with in control mice (Fig. 3.2a). Airway goblet cells increased to 38.0 % of total airway cells in OVA-treated mice compared with 0.4 % in saline control mice ($P = 0.0001$, OVA vs. saline; Fig. 3.4a). The mucus occlusion of the airway diameter morphometric score increased 15-fold in the OVA-treated mice compared with control mice ($P < 0.0001$, OVA vs. saline; Fig. 3.4b). Allergen-induced goblet cell hyperplasia and mucus occlusion of airway diameter were inhibited by both the (R)-enantiomer (Fig. 3.2c) and (S)-enantiomer (Fig. 3.2d) of albuterol. By means of morphometric analysis, (R)-albuterol decreased goblet cell hyperplasia by 48.9 % ($P = 0.0066$, R-albuterol/OVA vs. OVA; Fig. 3.4a) and airway mucus occlusion by 41.4 % ($P = 0.0042$, R-albuterol/OVA vs. OVA; Fig. 3.4b). (S)-Albuterol reduced goblet cell hyperplasia by 44.8 % ($P = 0.0095$, S-albuterol/OVA vs. OVA; Fig. 3.4a) and mucus occlusion of the airways by 35.7 % ($P = 0.0088$, S-albuterol/OVA vs. OVA; Fig. 3.4b).

Airway Edema. Airway edema was observed in the lungs of OVA-treated mice (Fig. 3.2b) compared with that seen in saline-treated control mice

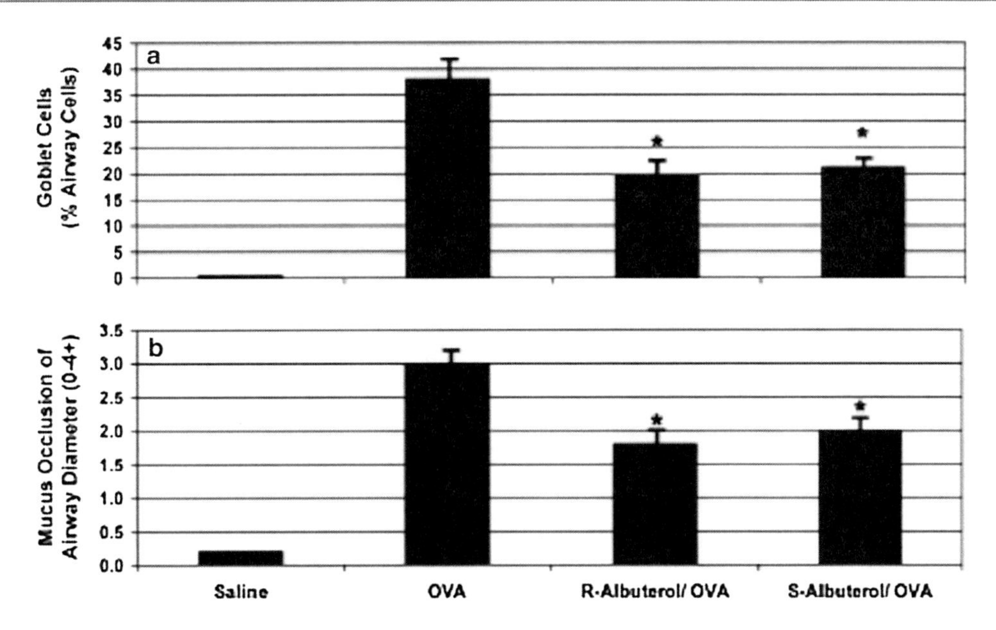

Fig. 3.4 Effect of (R)- and (S)-albuterol enantiomers on allergen-induced airway mucus hypersecretion. The number of goblet cells (**a**) and mucus occlusion of airway diameter (**b**) were determined. $^{*}P < 0.05$ versus OVA

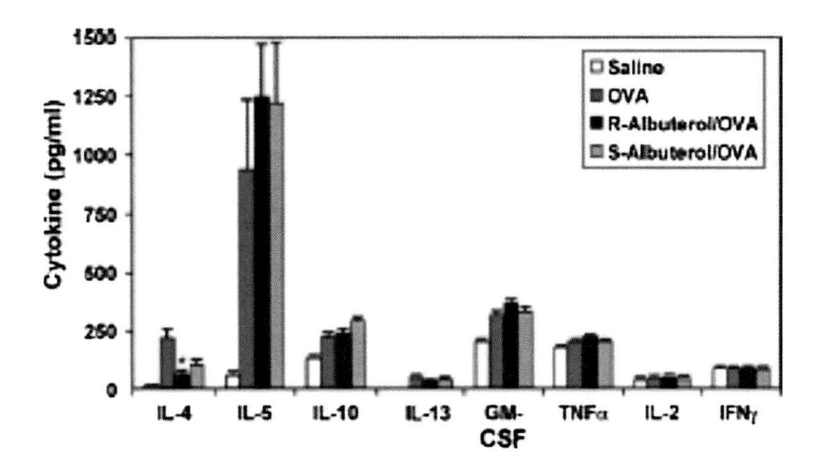

Fig. 3.5 Effect of (R)- and (S)-albuterol enantiomers on BAL fluid cytokine levels in OVA-treated mice. BAL fluid was assayed for T_H1 and T_H2 cytokines. $^{*}P < 0.05$ versus OVA

(Fig. 3.2a). (S)-Albuterol markedly increased airway edema in OVA-treated mice (Fig. 3.2d vs. b). In contrast, (R)-albuterol had no effect on allergen-induced edema in the airways of OVA-treated mice (Fig. 3.2c).

Cytokine Release

Significant levels ($P < 0.05$) of IL-4, IL-5, IL-13, and GM-CSF were found in the BAL fluid of OVA-treated mice compared with levels in the saline control group (Fig. 3.5). The increased levels of IL-4 in OVA-treated mice were reduced 70.5 % and 52.2 % by (R)-albuterol and (S)-albuterol, respectively; the reduction was statistically significant only for the (R)-enantiomer ($P = 0.043$, R-albuterol/OVA vs. OVA; Fig. 3.5). There was no significant effect of either the (R)- or (S)-enantiomer of albuterol on the increased BAL fluid levels of IL-5, IL-13, and GM-CSF in the OVA-treated animals. The levels of IL-10, TNF-α, IL-2, and IFN-γ were not significantly increased in the BAL fluid of OVA-treated mice compared with levels seen in saline-treated control mice; neither enantiomer affected the levels of these cytokines in OVA-treated animals (Fig. 3.5).

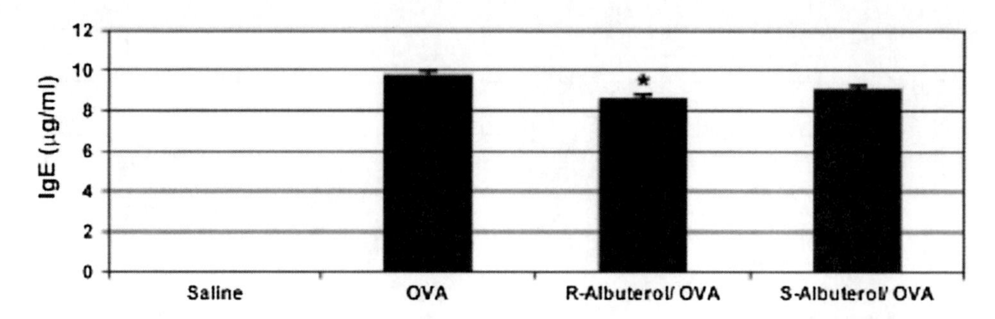

Fig. 3.6 (R)-Albuterol decreases OVA-specific IgE levels in OVA-treated mice. Plasma OVA-specific IgE levels were determined. $^{*}P < 0.05$ versus OVA

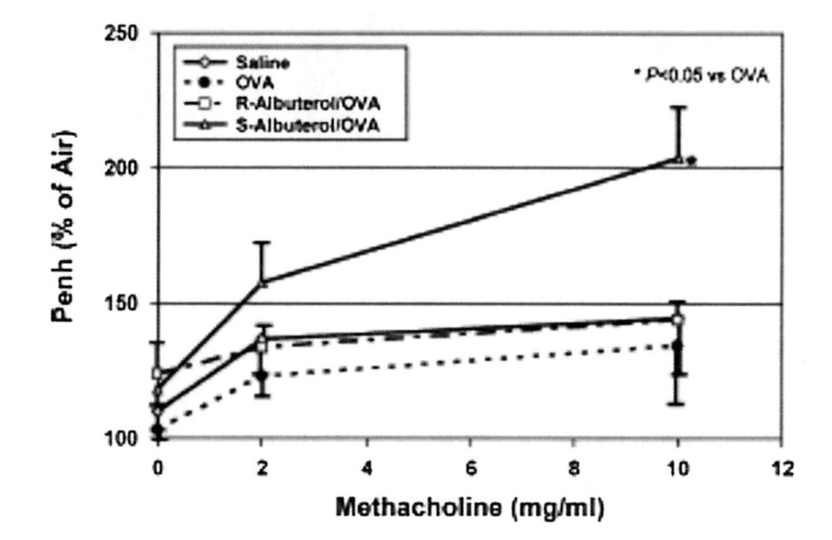

Fig. 3.7 (S)-Albuterol increases allergen-induced airway hyperresponsiveness. The degree of bronchoconstriction to aerosolized methacholine (0, 2, and 10 mg/mL) was expressed as P_{enh} (percentage of air as control). $^{*}P < 0.05$ versus OVA

OVA-Specific IgE

Plasma OVA-specific IgE was absent in saline-treated control mice and present in OVA-treated control mice (Fig. 3.6). (R)-Albuterol and (S)-albuterol reduced OVA-specific IgE levels in OVA-treated mice; the reduction was statistically significant only for the (R)-enantiomer ($P = 0.0120$, R-albuterol/OVA vs. OVA; Fig. 3.6).

Effect of (S)- and (R)-Enantiomers of Albuterol on Allergen-Induced Airway Hyperresponsiveness

Pulmonary mechanics were assessed in response to aerosolized methacholine by means of noninvasive in vivo plethysmography on day 36, 24 h after the last intranasal OVA challenge. The OVA-sensitized mice had been challenged with three intranasal doses of OVA, a protocol that we have previously shown to induce airway inflammation and mucus hypersecretion but that is suboptimal for inducing airway hyperresponsiveness [9, 10]. This protocol was used because an augmenting effect of (S)-albuterol on airway hyperresponsiveness could have been masked in our mouse asthma model protocol, in which bronchial hyperresponsiveness is achieved through administration of 4 intranasal doses of OVA in mice sensitized by two intraperitoneal OVA doses [9, 10]. In OVA-treated mice (S)-albuterol significantly increased bronchial responsiveness to methacholine (Fig. 3.7). In contrast, (R)-albuterol did not alter airway responsiveness to methacholine in OVA-sensitized/OVA-challenged mice (Fig. 3.7).

Effect of (R)- and (S)-Enantiomers of Albuterol in Non-OVA-Sensitized/OVA-Challenged Mice

Miniosmotic pumps containing either (R)- or (S)-albuterol were placed subcutaneously in saline control animals for a 24-day treatment period before pulmonary function testing and assessment of lung histopathology (Fig. 3.8) to examine the effect of the albuterol enantiomers independently of a modification of the allergic inflammatory response.

Lung Morphology. The (S)-enantiomer of albuterol did not induce airway edema independently of an allergic response. No airway edema or inflammation was seen in saline-treated mice administered either (S)-albuterol (Fig. 3.8b) or (R)-albuterol (Fig. 3.8c).

Pulmonary Mechanics

The (S)-enantiomer of albuterol did not affect airway reactivity in non-OVA-sensitized/ OVA-challenged mice. At 0, 2, and 10 mg/mL methacholine challenge doses, Penh (percentage of air) values were, respectively, 105.1 %, 92.3 %, and 100.0 % of the values of saline control animals. Similarly, (R)-albuterol had no effect on Penh (percentage of air) values of the saline control group. At 0, 2, and 10 mg/mL methacholine doses, non-OVA-sensitized/OVA-challenged mice administered (R)-albuterol had 106.4 %, 95.7 %, and 101.1 % Penh (percentage of air) values of the saline control animals.

Discussion

In this mouse model of asthma, we found both overlapping and distinct actions of the (S)- and (R)-enantiomers of albuterol on key features of allergen-induced airway inflammation and responsiveness to methacholine. (R)-Albuterol significantly reduced the following features of allergen-induced airway inflammation: BAL fluid levels of IL-4 and eosinophils, airway eosinophil infiltration, goblet cell hyperplasia, and mucus occlusion and circulating levels of OVA-specific IgE. Although (S)-albuterol also decreased airway tissue eosinophilia, goblet cell hyperplasia, and mucus plugging of the airways, no significant effect on the influx of eosinophils into the BAL fluid was observed. (S)-Albuterol, but not (R)-albuterol, had adverse effects on airway function by increasing airway edema and hyperresponsiveness in OVA-treated mice. We found in this mouse asthma model that both (S)- and (R)-albuterol inhibited infiltration of eosinophils into airway tissue, but only (R)-albuterol significantly reduced eosinophil influx into BAL fluid. Although a correlation between BAL fluid levels of IL-5 and eosinophilia is typically seen, there was no reduction in the increased IL-5 levels of OVA-treated mice administered either albuterol enantiomer. Short- and long-acting β2-adrenergic agonists might facilitate eosinophil apoptosis, thereby reducing airway eosinophilia [11]. Albuterol (0.1–10 mM) in vitro dose dependently decreases colony numbers and increases apoptosis of eosinophil progenitor cells from the blood of patients with asthma [12]. In contrast, racemic albuterol, (R)-albuterol, and (S)-albuterol do not affect apoptosis of antigen-specific human T-cell lines [3]. An unexpected finding of this study was the reduction in airway goblet cell hyperplasia and mucus hypersecretion by both the (R)- and (S)-enantiomers of albuterol. Limited data exist regarding the effect of (R)- and (S)-enantiomers of albuterol on airway mucus gland function. Sympathetic (i.e., adrenergic), parasympathetic (i.e., cholinergic), and sensory-efferent (i.e., tachykinin-mediated) pathways regulate mucus secretion from airway epithelial goblet cells and submucosal glands [13]. In human airways the cholinergic response is predominant and is mediated by muscarinic M3-receptors on the mucus secretory cells [13]. In ovine tracheal epithelial cells, (R)-albuterol increased ciliary beat frequency, whereas (S)-albuterol had no significant effect [14]. Increased mucociliary clearance rates by β2-adrenergic agonists have been reported in some patients with asthma [15]. The TH2 cytokines IL-4 and IL-13 have potent effects on mucus secretion [16, 17]. Administration of each cytokine independently

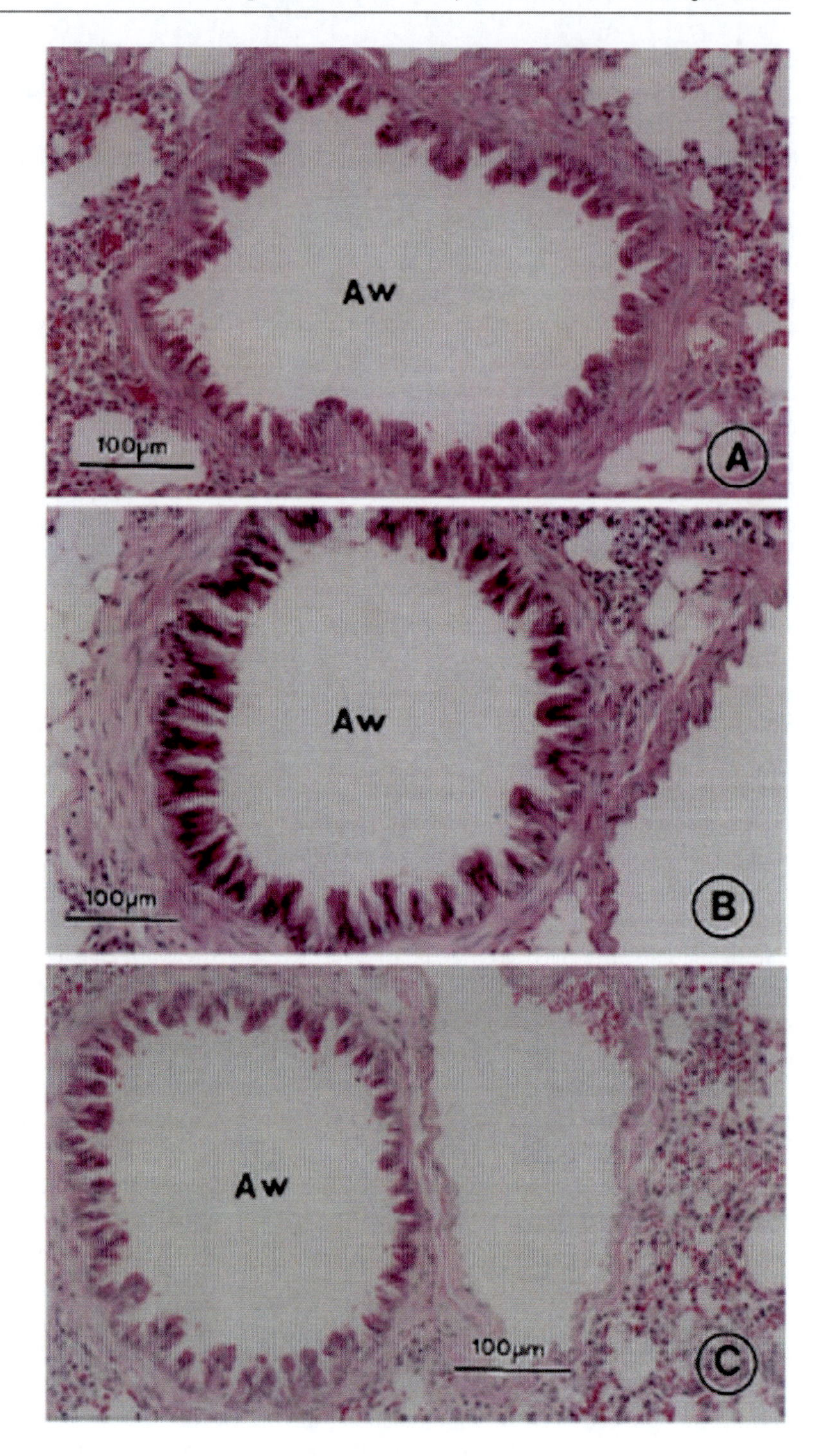

Fig. 3.8 Effect of (R)- and (S)-albuterol enantiomers on airway histology in non-OVA-sensitized/non-OVA-challenged mice. Lung tissue was obtained on day 36 from saline control animals (**a**), saline-treated mice administered (S)-albuterol (**b**), and saline-treated mice administered (R)-albuterol (**c**), and sections were stained with hematoxylin and eosin. *AW* airway. *Bars* = 100 μm

stimulates airway mucus accumulation in mice [17]. MUC5AC gene expression and BAL fluid mucus protein release are increased in IL-4 transgenic mice [16]. In addition, inhibition of IL-4 by administration of soluble IL-4 receptor reduces airway mucus hypersecretion and inflammatory cell trafficking to the lungs in OVA-treated mice [18]. We found that the increased BAL fluid levels of IL-4, but not IL-13, Fig. 3.5. Effect of (R)- and (S)-albuterol enantiomers on BAL fluid cytokine levels in OVA-treated mice. BAL fluid was assayed for TH1 and TH2 cytokines. $*P < 0.05$ versus OVA Fig. 3.4. Effect of (R)- and (S)-albuterol enantiomers on allergen-induced airway mucus hypersecretion. The number of goblet cells (A) and mucus occlusion of airway diameter (B) were determined. $*P < 0.05$ versus OVA. J ALLERGY CLIN IMMUNOL VOLUME 116, NUMBER 2 Henderson, Banerjee, and Chi 337 in OVA-treated mice were significantly reduced by (R)-albuterol, with a trend toward IL-4 reduction by (S)-albuterol. Thus the albuterol enantiomers might modulate allergen-induced airway inflammation and mucus hypersecretion through IL-4, rather than IL-5 or IL-13, signaling. We have recently demonstrated a similar discordance between IL-4 and IL-5/IL-13 in mediation of allergen-induced airway eosinophilia and mucus hypersecretion [19]. In a mouse asthma model, the selective redox effector factor 1 inhibitor PNRI-299, which inhibits the transcription factor activator protein-1 (AP-1), significantly decreased airway eosinophil infiltration and mucus occlusion and lung gene expression of IL-4 but not IL-5 or IL-13 [19]. During T-cell activation, a complex interaction between nuclear factor of activated T cells and AP-1 is necessary for inducible expression of IL-4 [20]. In contrast, transcriptional regulation of IL-5 and IL-13 might be independent of AP-1 binding [21]. The induction of goblet cell hyperplasia and eosinophilia by IL-4 in triple IL-5/IL-9/IL-13-knockout mice further demonstrates the key role IL-4 exerts in the development of the TH2 phenotype [22]. In our studies the reduction in IL-4 in OVA-treated mice administered (R)-albuterol correlated with a decrease in circulating OVA-specific IgE.

We found that (S)-albuterol, but not (R)-albuterol, augmented the interstitial edema observed in OVA-treated mice, suggesting a proinflammatory effect unique to (S)-albuterol. This effect of (S)-albuterol was not observed in nonsensitized/nonchallenged mice. In 2 sheep models of altered lung fluid balance, the effect of aerosolized racemic albuterol and its (R)- and (S)-enantiomers on lung epithelial permeability has been examined [23]. Pretreatment with (S)-albuterol increased the level of albumin in the epithelial lining fluid in sheep receiving histamine to increase lung permeability. This effect was not observed after pretreatment with either (R)-albuterol or its racemate. In sheep receiving an increase in left atrial pressure to increase hydrostatic forces, only (S)-albuterol increased lung water volume. Thus, in the presence of altered lung fluid balance, (S)-albuterol, but not (R)-albuterol, might increase lung epithelial permeability. Our finding that (S)-albuterol increases allergen-induced interstitial edema indicates a potential adverse effect of this enantiomer in patients with asthma. (S)-Albuterol significantly increased bronchial responsiveness to methacholine challenge in OVA-sensitized/OVA-challenged mice, an adverse effect on airway function not shared by (R)-albuterol. In nonsensitized/nonchallenged mice, (S)-albuterol did not independently affect the airway response to methacholine. We used whole-body plethysmography to assess airway hyperreactivity to methacholine in the study groups. Although there has been recent controversy regarding the use of Penh as an indirect measure of pulmonary mechanics [24–26], Penh values correlate well with airway resistance measured directly in anesthetized, tracheotomized, and mechanically ventilated mice [27, 28]. A strong correlation also exists between Penh values and the intensity of the allergen-induced airway eosinophil infiltration in the mouse asthma model [29]. (S)-Albuterol might mediate its augmenting effect on bronchial hyperresponsiveness through several modes of action, including parasympathetic and sympathetic pathways. In OVA-sensitized guinea pigs, continuous exposure to (R,S)- and (S)-albuterol, but not (R)-albuterol, for a 10-day period increased bronchial hyperresponsiveness to both histamine and OVA [5]. Chronic capsaicin treatment prevented the (R,S)- and (S)-albuterol-induced bronchial hyperresponsiveness in this model to indicate the importance of capsaicin-sensitive sensory nerves in (S)-albuterol-mediated development of airway hyperresponsiveness [5]. Recent studies by Agrawal et al. [4] indicate that (S)-albuterol activates proconstrictor and proinflammatory pathways in human bronchial smooth muscle cells. (S)-Albuterol significantly increased the expression and activity of Gia-1 protein and reduced Gs protein in these cells [4]. (S)-Albuterol also increased the intracellular free-calcium concentration in the bronchial smooth muscle cells after methacholine stimulation [4]. These proconstrictor effects of (S)-albuterol were accompanied by stimulation of phosphatidylinositol 3-OH-kinase and nuclear factor kB proinflammatory pathways in the smooth muscle cells. (R)-Albuterol induced the opposite effects on Gia-1, Gs, and intracellular free-calcium concentration in the bronchial smooth muscle cells, which indicates separate mechanisms of action of the enantiomers.

In summary, the actions of the (R)- and (S)-enantiomers of albuterol in the lungs of allergen-sensitized/allergen-challenged mice are complex. Although both enantiomers reduce mucus hypersecretion and trafficking of eosinophils to the lungs after allergen challenge, only the (S)-enantiomer induces airway edema and hyperresponsiveness to methacholine. Additional studies are needed to delineate the specific effects of the (R)- and (S)- enantiomers of albuterol on airway inflammation and hyperresponsiveness in patients with asthma.

Conclusion

Both (R)- and (S)-enantiomers of albuterol reduce airway eosinophil trafficking and mucus hypersecretion in a mouse model of asthma. However, (S)-albuterol increases allergen-induced airway edema and hyperresponsiveness. These adverse effects of the (S)-enantiomer on lung function might limit the clinical efficacy of racemic albuterol.

Materials and Methods Used in the Study

Study Protocol

All animal use procedures were approved by the University of Washington Animal Care Committee. Female BALB/c mice (6–8 weeks of age; The Jackson Laboratory, Bar Harbor, Me) received an intraperitoneal injection of 100 mg of ovalbumin (OVA; 0.2 mL of 500 mg/mL) complexed with alum on days 0 and 14 (Fig. 3.1). Mice were anesthetized with 0.2 mL of ketamine (6.5 mg/mL)/xylazine (0.44 mg/mL) in normal saline administered intraperitoneally before receiving an intranasal dose of 50 mg of OVA (50 mL of 1 mg/mL) on days 14, 25, and 35 (Fig. 3.1). The control group received 0.2 mL of normal saline with alum administered intraperitoneally on days 0 and 14 and 0.4 mL of saline without alum administered intranasally on days 14, 25, and 35. In both the saline- and OVA-treated groups, miniosmotic pumps (200 mL, Alzet Model 2004; Durect Corp, Cupertino, Calif) containing either (R)- or (S)-albuterol (1 mg _ kg21 _ d21, 6 mL/day delivery administration) were inserted subcutaneously on the back slightly posterior to the scapulae on day 13 and remained in place until study conclusion on day 36 (Fig. 3.1). Absorption of the compounds by local capillaries results in systemic administration. Each study group consisted of 4–6 animals. The 1 mg _ kg21 _ d21 dose of albuterol enantiomer infusion was selected on the basis of prior work by Sartori et al [30], demonstrating that continuous release of racemic albuterol (2 mg _ kg21 _ d21) subcutaneously by means of miniosmotic pump produced steady-state, high-plasma levels of albuterol (1,025 M) in mice.

Pulmonary Function Testing

In vivo airway responsiveness to methacholine was determined on day 36 in conscious, free-moving, spontaneously breathing mice by using whole-body plethysmography (Model PLY 3211; Buxco Electronics Inc, Sharon, Conn), as described by Hamelmann et al. [27]. Mice were challenged with aerosolized saline or increasing doses (2 and 10 mg/mL) of methacholine generated by an ultrasonic nebulizer (DeVilbiss Health Care, Inc, Somerset, Pa) for 2 min. The degree of bronchoconstriction was expressed as enhanced pause (Penh), a calculated dimensionless value that correlates with measurement of airway resistance, impedance, and intrapleural pressure. Penh readings were taken and averaged for 4 min after each nebulization challenge. Penh is calculated as follows: Penh

¼ ½ðTe = Tr21 + 3ðPEF = PIF + _, where Te is expiration time, Tr is relaxation time, PEF is peak expiratory flow, and PIF is peak inspiratory flow 3 0.67 coefficient. The time for the box pressure to change from a maximum to a user-defined percentage of the maximum represents the relaxation time. The Tr measurement begins at the maximum box pressure and ends at 40 %. Because Penh is the ratio of measurements obtained during the same breath, it is mainly independent of functional residual capacity, tidal volume, and respiratory rate.

Light Microscopy-Morphometry

After pulmonary function testing, bronchoalveolar lavage (BAL) was performed on the right lung, with total BAL fluid cells counted and eosinophils identified by means of eosin staining. Left lung tissue was obtained for histopathology, and plasma was obtained for OVA-specific IgE levels. Ten lung sections per animal were randomly selected and examined in a blinded manner. Sections were stained with hematoxylin and eosin, the total inflammatory cell infiltrate was assessed on a semiquantitative scale (0–41), the number of eosinophils per unit of airway area (2,200 mm^2) was determined by using a point-counting system (Image-Pro Plus point-counting system software, Version 1.2 for Windows; Media Cybernetics, Silver Spring, Md) [31], and interstitial and perivascular airway edemas were assessed [19, 32].

Airway Goblet Cells

As a percentage of total airway cells were identified by means of Alcian blue staining, the degree of mucus plugging of the airways (0.5–0.8 mm in diameter) with the percentage occlusion of the airway diameter was classified on a 0–41 scale on the basis of the following criteria: 0, no mucus; 1, approximately 10 % occlusion; 2, approximately 30 % occlusion; 3, approximately 50 % occlusion; and 4, greater than approximately 80 % occlusion.

Cytokine Assays

IL-4, IL-5, IL-10, GM-CSF, TNF-α, IL-2, and IFN-γ were assayed in BAL fluid with Bio-Plex Mouse Cytokine assays (Bio-Rad Laboratories, Hercules, Calif) that are bead-based multiplex sandwich immunoassays with a limit of detection of less than 10 pg/mL. IL-13 was assayed in BAL fluid with a mouse IL-13 immunoassay (Quantikine M; R&D Systems, Minneapolis, Minn), with a limit of detection of less than 1.5 pg/mL.

OVA-Specific IgE Assay

OVA-specific IgE was assayed by modification of the method of Iio et al. [33]. Nunc 96-well flat-bottom plates (Nalge Nunc International, Rochester, NY) were coated with 50 mg/mL OVA in 13 PBS overnight at room temperature, washed 3 times with 13 PBS plus 0.05 % Tween-20 (wash buffer), blocked with 3 % BSA in 13 PBS for 1 h at room temperature, and washed 4 times with wash buffer. Fifty-microliter plasma samples (1:1 in 13 PBS) were added per well and incubated for 90 min at 37_C, then washed 4 times with wash buffer, and blotted dry by inverting over paper towels. One hundred microliters (1:100 in 13 PBS) of biotin-conjugated rat anti-mouse IgE mAb (clone R35-72; BD Biosciences, San Diego, Calif) was added to each well and incubated overnight at 4_C and then washed 4 times with wash buffer and blotted dry. Then, 100 mL per well (1:1,000 in 13 PBS) streptavidin-horseradish peroxidase-conjugated secondary antibody (BD Biosciences) was added, and samples were incubated at 37_C for 90 min, then washed 4 times with wash buffer, and blotted dry. One hundred microliters of substrate solution (i.e., 1 tablet of 2,2-azinobis [3-ethylbenzthiazoline-sulfonic acid, ABTS; Sigma Chemical Co, St Louis, Mo] dissolved in 100 mL of 0.05 M phosphate-citrate buffer, pH 5.0, and 25 mL of 30 % H_2O_2) was added per well, and color was developed for 30 min at room temperature. OD 405 nm was measured by using an AD 340C absorbance detector (Beckman Coulter Inc, Fullerton, Calif). For the IgE standard curve, a sandwich ELISA was used in which in separate assay plates biotin-conjugated rat anti-mouse IgE mAb (clone R35-72, BD Biosciences) was used to coat the wells and, instead of plasma samples, known concentrations of purified anti-mouse IgE (clone C38-2, BD Biosciences).

Research Area 2. Studies on Prophylactic and Therapeutic Strategies to Combat Some Local and Systemic Inflammatory Pathologies

Overall Objective of This Series of Studies

The twin banes of inflammation and degeneration of functional cells are the main issue in diseases such as asthma and other pulmonary and systemic inflammation; the critical tissues are first infested with aggressive populace from the pool of circulating immune cells that, along with the mediators they sequester and secrete locally, throw the homeostasis of the tissues out of gear. The end result is that while unregulated inflammation forms plaques of inflammatory cells and lesions in the inflamed tissue, the relevant cells slowly die out and are never replaced by the body as due to the unregulated depletion due to inflammation; tissue resident stem cells are unable to replace lost cells at the same rate or not at all. This results in denudation of function of the tissue and body as a whole leading to multiple organopathies that completely incapacitate the individual.

While pharmacological intervention manages the disease and minimizes the discomfort of the afflicted person, regeneration of lost healthy functional tissue is key to actual disease modification. With this view the following series of studies were conducted in order to understand the pattern of cellular traffic and zoom in on the cluster of associated molecules that may together or sequentially perpetrate the above malfunctions.

The approaches were as follows:

(a) Understand basic biology of the disease to dissect per (i) onset, (ii) establishment, (iii) maintenance, and (iv) exacerbation of the disease and study fundamental disease etiology.
(b) Identify and characterize key cells and mediators (i.e., *functional targets*) critical for the correct execution of such pathways through the "omics" circuit completion.
(c) Provide alternative solution for significant disease modification and validation in both cell-based (cell line or primary cell-based) models in vitro or ex vivo as well as in vivo preclinical models using mouse models via:
    (I) Pharmaceutical small molecules
    (II) Alteration of nutraceutical milieu in 2D cell culture and tissue engineering

Major Findings and Their Interpretations Under Research Area 2

Subchapter 1: Role of Integrin α4 (VLA – Very Late Antigen 4) and Integrin β2 (CD18) in a Pulmonary Inflammatory and a Systemic Disease Model Using Genetic Knockout Mice

Role of Integrins α4 and β2 in Onset and Development of Acute Allergic Asthma in Mice

Summary of the Study

Objective. Recruitment of effector cell subsets to inflammatory lung, together with airway resident cells responsive to secreted products, plays pivotal roles in developing and maintaining asthma. Differential use of adhesion molecules dictates the recruitment patterns of specific cell subsets; yet a clear understanding of the distinctive adhesive molecular pathways guiding them to the lung is lacking. To provide further insight into the role of α4b1/VCAM-1 pathway and to compare this to the role of β2 integrin in the development of acute asthma phenotype, we used genetically deficient mice, in contrast to previous studies with anti-functional antibodies yielding ambiguous results.

Methods. Allergen-dependent airway inflammation and hyperresponsiveness was induced in conditional α4Δ/Δ, VCAM-1 L/L, and β2 L/L mice. Cytology, immunocytochemistry, cytokine and immunoglobulin measurements, and cell type accumulation in the lung, BAL fluid, plasma, and hematopoietic tissues were carried out.

Results. Asthma phenotype was totally abrogated in α4- or β2-deficient mice. Adoptive transfer of sensitized α4Δ/Δ CD4+ cells into challenged normal mice failed to induce asthma, whereas α4+/+ CD4+ cells were able to induce asthma in challenged α4Δ/Δ mice. Parallel studies with β2 L/L or VCAM-1 L/L mice uncovered novel mechanistic insights in primary sensitization and into redundant or unique functional roles of these adhesion pathways in allergic asthma.

Conclusions. The lack of α4 integrin not only impedes the migration of all white cell subsets to lung and airways but also prevents upregulation of vascular cell adhesion molecule-1 (VCAM-1) in inflamed lung vasculature and, unlike β2, attenuates optimal sensitization and ovalbumin-specific IgE production in vivo. As VCAM-1 deficiency did not protect mice from asthma, interactions of α4b1+ or α4b7+ cells with other ligands are suggested.

Background and Objective of the Study

Asthma is a chronic allergic airway disease characterized by persistent inflammation and airway hyperresponsiveness (AHR). T cells, especially Th2 cells secreting IL-4, IL-5, and IL-13, are pivotal in orchestrating the disease process, and adoptive transfer of Ag-primed T cells in naïve animals can induce eosinophilia, AHR, and late airway responses [34]. Apart from Th2, other effector cells for AHR and asthma are the eosinophils and mast cells. Under the influence of chemoattractants and IL-5, eosinophils proliferate, migrate into the lung, and are activated to secrete histamine, leukotrienes, and other mediators [35]. Mature mast cells are sources of histamine, proteases, heparin, and lipid mediators, released upon allergen-induced IgE cross-linking as the initial step in the inflammation process. Human studies of allergic asthma and animal models of asthma continue to provide conflicting data on the contribution of airway resident cells versus specific cell subsets recruited from circulation in the development of asthma phenotype [34, 36, 37]. The preferential recruitment of Th2 cells and eosinophils [37] in the lung is mediated by a cascade of adhesive interactions initiated by activated endothelial cells attracted by locally elaborated chemokines/cytokines. However, the distinct molecular pathways that dominate these processes are continuously under revision. A key question is whether hematopoietic cells from circulation (i.e., eosinophils, neutrophils) or resident cells (i.e., tissue macrophages) play a pivotal role in orchestrating the acute allergic asthma phenotype and the role that adhesion molecules play in the interface of such interaction. α4 (very late activation antigen-4 [VLA-4] or α4b1) [38] and β2 integrins (lymphocyte function-associated antigen [LFA-1] or aLβ2; Mac-1 or aMβ2) [39] mediate hematopoietic cell migration in various inflammatory diseases [40, 41]. VLA-4 interacts with extracellular matrix components (fibronectin, osteopontin) and with vascular cell adhesion molecule-1 (VCAM-1) mediating adhesion to activated endothelium. It also interacts with junctional adhesion molecule-2 (JAM-2), with mucosal addressin-associated cell adhesion molecule-1 (MAdCAM-1), and with itself, potentially facilitating interactions among leukocytes in inflammation [42]. Studies on expression of adhesion molecules in migrated cells have not provided definitive answers. Other than the CD18_/_ [43] mouse, in vivo and in vitro studies thus far have employed only anti-functional antibodies to several adhesion molecules (i.e., VLA-4 and LFA-1). These studies yielded conflicting data depending on the type and dose of anti-α4 antibody used, its route of administration, or the animal model used [44–51] and examined effects after primary sensitization.

To gain further insight into the contribution of α4 integrin to asthma phenotype and further refine previous results from studies with anti-α4 antibodies, we used conditionally ablated α4 integrin or VCAM-1 mice subjected to ovalbumin (OVA)-induced and Th2 cytokine-driven lung allergic inflammation. Our data examining sensitization issues in the absence of α4 integrin yield novel insights into the involvement of α4 in this process, and a comparison of the data to β2 knockout (ko) or VCAM-1-deficient mice provides a substantial refinement on their redundant and nonredundant contributions in asthma development.

Methods

Mice

Animal procedures were approved by the Institutional Animal Care Committee of the University of Washington. Mice (C57Bl/6.B129 background) used were Mxcre_α4f/f and Mx-cre + α4Δ/Δ (α4flox/flox bred to Mxcre + [52]). Ablation of α4 integrin in Mxcre + α4f/f mice was induced by 3 i.p. (intraperitoneal) injections of poly(I)-poly(C) (Sigma, St. Louis, MO, USA) during the first week of life. Neonatally ablated α4Δ/Δ Mxcre mice were studied as adults. β2 ko (CD18_/_) mice were previously described [43]. VCAM-1f/f mice were bred to Tie2cre + transgenics to generate VCAM-1-deficient mice as described [53]. In addition, in selected experiments, α4f/f mice also bred to Tie2cre transgenics to generate cre recombination of the α4 gene under Tie2 promoter as described [54] were used.

Induction of Acute Asthma Phenotype

Mice were immunized with 100 mg OVA (Pierce Chemical Co., Rockford, IL, USA) and complexed with aluminum sulfate (alum, Sigma, [9]) intraperitoneally injected on day 0. On day 8 (using 250 mg of OVA) and on days 15, 18, and 21 (using 125 mg of OVA), mice were anesthetized and given OVA by intratracheal (i.t.) administration [55]. Control groups received normal saline (complexed with alum) using the same schedule.

Pulmonary Function Testing

Twenty-four hours after the last intratracheal allergen (OVA) challenge (day 22), invasive pulmonary mechanics were measured in mice in response to methacholine as previously described [9].

Noninvasive Plethysmography

Allergen-induced hyperreactivity in OVA-treated mice was measured on day 22 by pulmonary-function responses to increasing doses of methacholine by noninvasive plethysmography [55] in conscious, free-moving, spontaneously breathing mice. The degree of bronchoconstriction was expressed as enhanced pause (Penh), a calculated dimensionless value that correlates with measurement of airway resistance, impedance, and intrapleural pressure.

Preparation of Cell Suspensions

Bone marrow cells were flushed from femurs [56]. Mice were bled retro-orbitally and red cells hemolyzed as described [56]. Bronchoalveolar lavage fluid (BALf) was collected as described [55]. Lung parenchymal cells (LP) were collected from lavaged lungs that were minced and digested as described in [46] and then vortexed and filtered through sterile 40-mm Nitex filter (Sefar America, Depew, NY, USA) before use.

FACS Analysis

Nucleated cells from peripheral blood (PB), bone marrow (BM), BAL fluid, and LP were analyzed on a FACSCalibur (BD Immunocytometry Systems, San Jose, CA, USA) using CellQuest software [52]. BD Biosciences (San Diego, CA, USA) antibodies used for cell-surface staining included CD45 (30 F-11), CD3 (145-2C11), CD4 (RM4-5), CD45RC (DNL-1.9), CD8 (53–6.7), β220 (RA3-6β2), IgM, CD19 (ID3), CD21 (7G6), CD23 (B3B4), GR-1 (RB6-8C5), and Mac1 (M1/70). F4/80 (Cl:A3-1(F4/80)) was obtained from Serotec Ltd. (Raleigh, NC, USA), and anti-α4 integrin (PS2) and anti-VCAM-1 (M/K-2) were from Southern Biotechnology (Birmingham, AL, USA). Irrelevant isotype-matched antibodies (BD Biosciences) were used as controls.

Smear Evaluation

Proportions of eosinophils and mast cells were assessed in Wright-Giemsa-stained smears. Histology and immunocytochemistry hematoxylin and eosin (H&E), Masson's trichrome stains, Alcian blue, and periodic acid Schiff staining were described [18]. Lung tissue sections were stained using anti-VCAM-1 (M/K-2) antibody [53].

Goblet Cell Metaplasia

Using a microscope, percentages of metaplastic goblet cells (detected by Alcian blue staining) in lung sections were calculated with counts from 10 different fields per mouse as described [9].

ELISPOT

IL-4+ and IFN-γ+ cells in single-cell suspensions from lung parenchyma and BALf were detected employing standard ELISPOT assays [43] using detection and capture monoclonal antibodies and AEC substrate reagent from BD Biosciences. Statistics data were analyzed using two-tailed Student's t-test, and p values less than 0.05 were considered statistically significant. Dots were counted manually using 40 × magnification.

Measurement of Cytokines

Th1/Th2 cytokines in BALf and plasma were assayed with mouse cytometric bead array (CBA, BD Biosciences). IL-13 and eotaxin were measured by Quantikine M kits from R&D Systems (Minneapolis, MN, USA).

OVA-Specific IgE and IgG1 in Plasma

Anti-mouse IgE (R35-72) and IgG1 (A85-1) antibodies (BD Biosciences) were used to measure OVA-specific IgE and IgG1 in plasma using sandwich ELISA [33, 57].

Adoptive Transfer

At day 8 after i.p. sensitization, 5×10^6 CD4+ splenocytes from both α4+/+ controls (α4f/fcre_) or α4Δ/Δ mice were purified by magnetic-activated cell sorting (MACS, Miltenyi Biotec, Auburn, CA, USA) and then injected into the tail veins of naïve controls or α4Δ/Δ recipients. The mice were subsequently challenged with 3 i.t. instillations of OVA over the next 72 h and sacrificed 24 h after the last instillation [58]. One group of α4Δ/Δ mice also received unpurified α4+/+ splenocytes from sensitized mice.

CFU-C Assay

To quantitate committed hematopoietic progenitors of all lineages, CFU-C assays were performed as described [52].

Proliferation Assay

MACS-separated CD4+ and CD8+ T cells from spleens were stimulated in vitro with various concentrations of stimuli (CD3/CD28, phorbol myristic acetate (PMA)/ionomycin, irradiated antigen-presenting cells (APCs), and lipopolysaccharide (LPS)) to assay proliferative responses. After 72 h, proliferation was measured either by CellTiter96 assay from Promega (Madison, WI, USA) measuring OD at 570 nm or by [3H]-thymidine incorporation following standard protocols.

Results in a Nutshell

Asthma phenotype was totally abrogated in α4- or β2-deficient mice. Adoptive transfer of sensitized α4Δ/Δ CD4+ cells into challenged normal mice failed to induce asthma, whereas α4+/+ CD4+ cells were able to induce asthma in challenged α4Δ/Δ mice. Parallel studies with β2−/− or VCAM-1−/− mice uncovered novel mechanistic insights in primary sensitization and into redundant or unique functional roles of these adhesion pathways in allergic asthma.

Detailed Results

α4Δ/Δ Mice Fail to Develop the Composite Asthma Phenotype in Response to Acute Allergen Challenge

Following initial sensitization and subsequent OVA challenges, control (α4f/fMxcre^{-}) mice develop the characteristic hallmarks of acute asthma, including eosinophil and lymphocyte sequestration in the lung and BALf, mucus hypersecretion, goblet cell metaplasia, and AHR to methacholine measured by either invasive or noninvasive plethysmography. None of the above changes were noted in our α4Δ/Δ mice (Figs. 3.9 and 3.10). Cells migrating to lung interstitium and airway lumen were quantified and their phenotypes analyzed by FACS (Fig. 3.10a) and cytospin smears. In OVA-treated controls, cells recovered from BALf were tenfold over alum-treated, but in α4Δ/Δ mice were unchanged from alum-treated mice (Fig. 3.10a). Likewise, fewer cells were recovered from lung parenchyma in OVA-treated α4Δ/Δ mice compared to similarly treated controls (Fig. 3.10a). If cells migrating to the lung or BALf are expressed as a fraction of

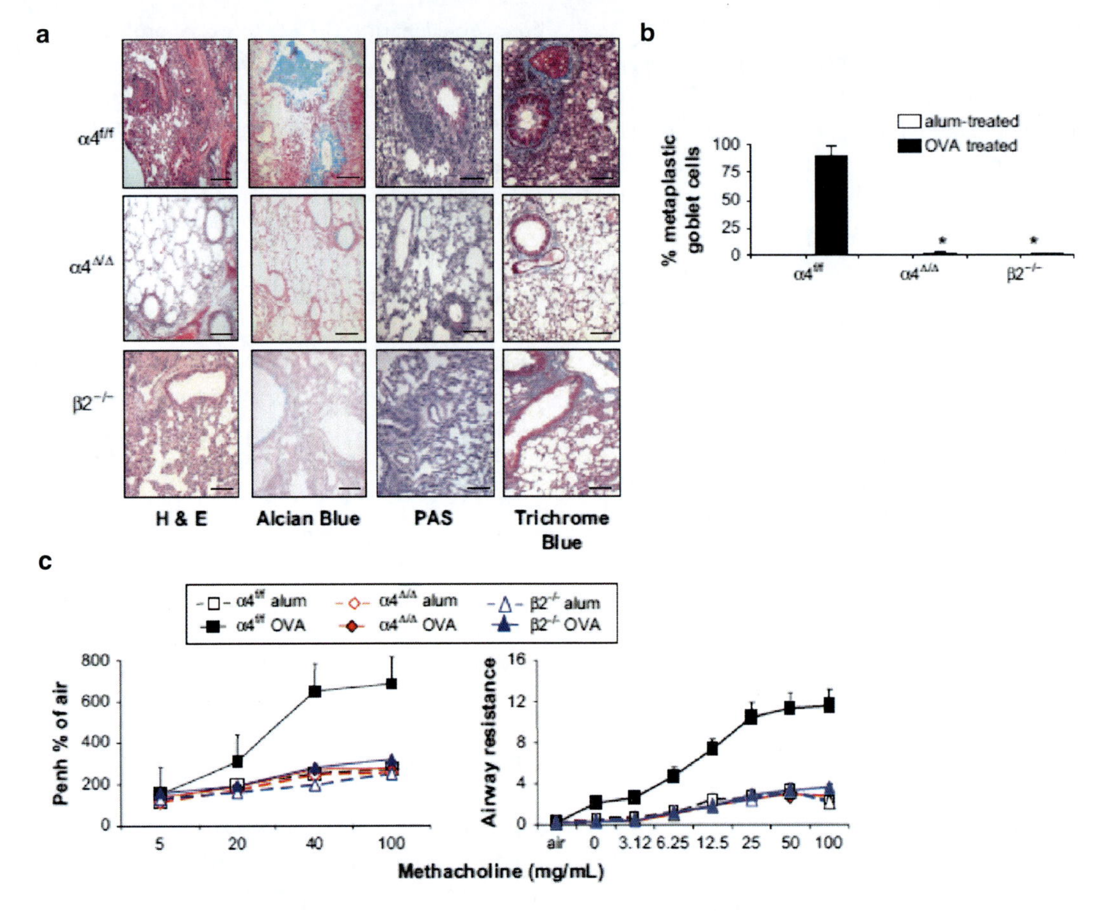

Fig. 3.9 Deletion of α4 and β2 integrins abrogates allergic lung disease in mice. Asthma phenotype was not induced in knockout mice as shown by the following: (**a**) Paraffin sections of lung tissue (*H&E*, *Alcian blue*, *PAS*, and *Masson's trichrome*). (**b**)% mean ± SEM of goblet cells [27] with metaplasia (*Alcian blue*), $^*P < 0.05$ comparing lung tissue from both $\alpha4^{\Delta/\Delta}$ and $\beta2^{-/-}$ mice to OVA-treated control lung. (**c**) In vivo airway responses to methacholine measure by whole-body plethysmography. Left panel: OVA-treated integrin-deficient mice do not show increased AHR to increasing doses of methacholine as control cre-$\alpha4^{f/f}$ littermates mice do ($p < 0.001$ compared to Penh values in alum-treated control and treated and alum-treated $\alpha4^{\Delta/\Delta}$ and $\beta2^{-/-}$ mice). Right panel: Invasive plethysmography of acute asthma in the same groups of mice as in C, except that $\alpha4^{\Delta/\Delta}$ were from Tie2cre$^+$ $\alpha4^{f/f}$ double-transgenic mice [21]

total circulating leukocytes (recruitment index), they represent 3 % in the lung and about 1 % in BALf (Table 3.1) in OVA-treated α4Δ/Δ mice, but 29 % and 33 % respectively in OVA-treated controls. Eosinophils, whose role in allergic asthma remains controversial [59, 60], were 18 % and 41 % of total white blood cells (WBCs) recovered from the lung and BALf respectively in controls, but only 2 % and 1 % in α4Δ/Δ mice (Fig. 3.10a). Further, IL-4+ cells were low in LP and BALf, and IFN-γ+ cells were low in BALf in α4Δ/Δ mice (Fig. 3.10b).

To explore whether hematopoietic progenitor cells (HPCs), which are found to be increased post-OVA in circulation [61], migrate to lung or BALf along with mature cells, we quantitatively assessed them by clonogenic CFU-C assays. We found increased progenitor cells in both lung and BALf in control, but not in α4Δ/Δ, mice (Fig. 3.11), despite the higher circulating HPC levels in the latter. Thus, the migratory behavior of progenitor cells to lung followed a pattern similar to that of mature cells in all mice.

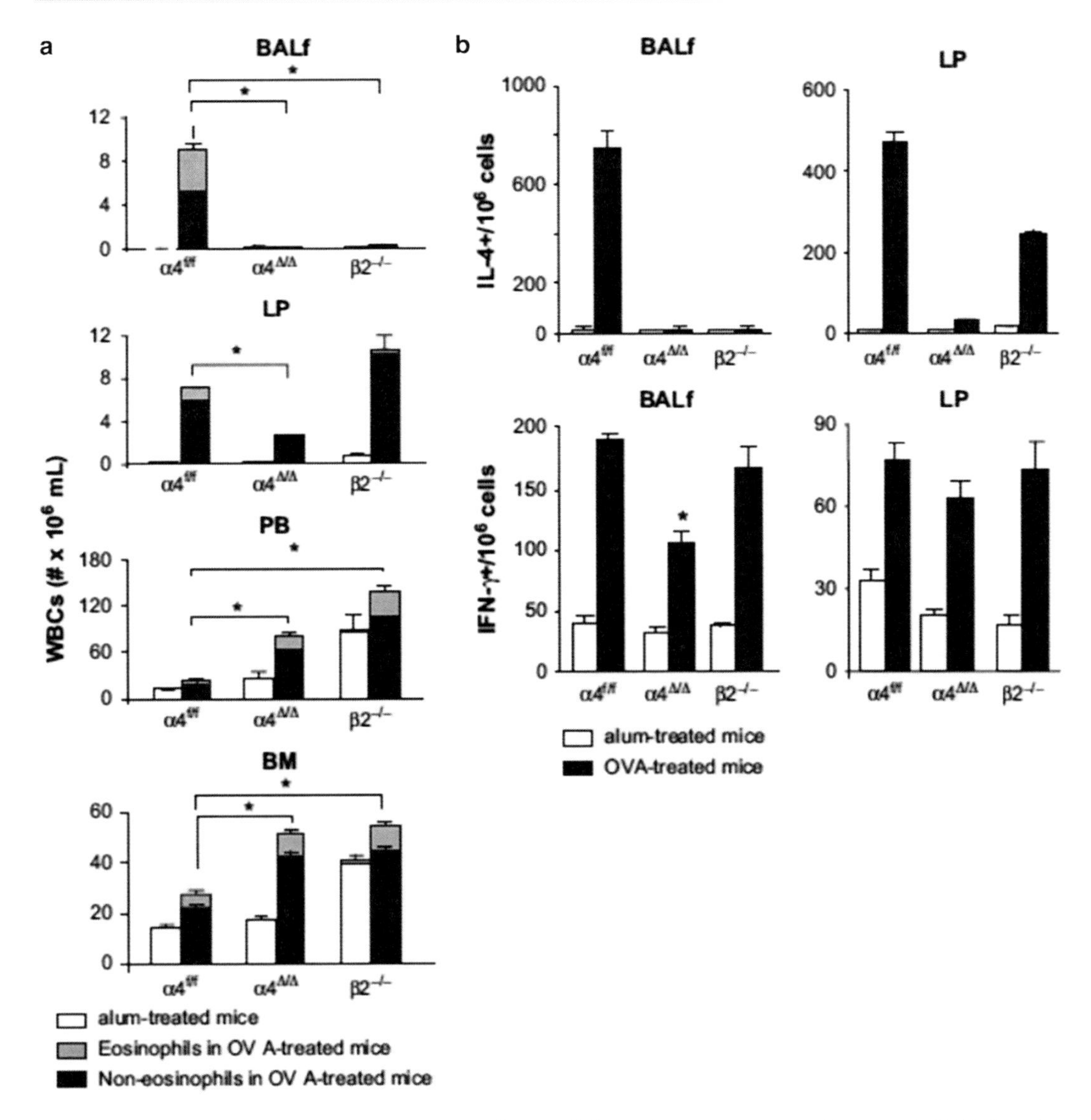

Fig. 3.10 Local and systemic cellular and humoral responses to OVA challenge in control and OVA-treated mice. (**a**) Numbers of total WBCs in alum-treated mice (*white bars*) and OVA-treated mice (*black/gray bars*) are shown. Height of *gray bars* denotes number of eosinophils and *black bars* represent noneosinophils in BAL from both lungs, LP from both lungs, peripheral blood and bone marrow from both femurs, $^{*}P < 0.05$ post-OVA in integrin-deficient mice compared to post-OVA control. Cell numbers are represented as men $\pm$ SEM, 12 mice were tested per group. (**b**) *Top panel*: numbers of IL-4^{+} cells in 10^{6} cells from BALf and LP as detected by ELISPOT from alum-treated mice (*white bars*) and OVA-treated mice (*black bars*). *Lower panel*: numbers of IFN-γ^{+} cells in 10^{6} cells from BALf and LP in the same mice, $n = 12$/group, $^{*}P = 0.05$ post-OVA in integrin-deficient mice compared to post-OVA control

OVA-Specific Sensitization Responses in α4Δ/Δ Mice

To examine whether α4Δ/Δ mice responded optimally to OVA sensitization, we evaluated the elaboration of OVA-specific IgE in these mice. OVA-specific IgG1 was similar in all three genotypes, but OVA-specific IgE was significantly lower only in α4Δ/Δ mice (Fig. 3.12a). Since the class switch from IgG1 to IgE is largely Th2 cytokine dependent [62], we measured the systemic elaboration of Th1-dependent (IL-2, IFN-γ, TNF-α) and Th2-dependent (IL-4, IL-5, IL-13)

Table 3.1 Cellular responses of leukocyte subsets to OVA treatment in all groups of mice studied

Blood								
	Total	Lymphocytes						
	WBCs	CD3+	β220+	Monocytes	Neutrophils	Eosinophils	Basophils	
WT	25.92 ± 6.41	7.96 ± 2.45	6.25 ± 1.07	1.7 ± 2.48	8.38 ± 2.05	0.79 ± 1.56	1.11 ± 0.03	
$\alpha4^{\hat{a}\acute{I}LE/\hat{a}\acute{I}LE}$	79.45* ± 22.24	26.3* ± 1.18	19.6* ± 3.75	5.72 ± 1.18	21.42* ± 4.85	6.37* ± 2.05	0.04* ± 0.001	
$B2^{-/-}$	147.28* ± 60.16	43.3* ± 8.22	14.6* ± 1.53	8.41* ± 0.13	57.87* ± 3.97	22.76* ± 2.73	0.34* ± 0.01	
$VCAM\text{-}1^{-/-}$	18.79 ± 3.58	5.108 ± 1.97	4.83 ± 1.35	1.72 ± 0.89	0.65 ± 0.07	0.875 ± 0.01	0.875 ± 0.01	
Lung								
	Total	Lymphocytes					Basophils	
	WBCs	CD3+	β220+	Monocytes	Neutrophils	Eosinophils	Mast	Mφ
WT	7.18 ± 0.12	2.84 ± 0.05	0.06 ± 0.01	1.02 ± 0.03	1.36 ± 0.01	1.3 ± 0.002	0.028 ± 0.001	1.8 ± 0.07
$\alpha4^{\hat{a}\acute{I}LE/\hat{a}\acute{I}LE}$	2.64* ± 0.04	0.42* ± 0.07	0.0076* ± 0.0001	0.592* ± 0.04	1.238 ± 0.02	0.06* ± 0.001	0	0.36* ± 0.002
$B2^{-/-}$	10.7 ± 0.97	0.008* ± 0.02	0.2 ± 0.001	1.66 ± 0.05	3.96* ± 0.06	0.58* ± 0.01	0	3.4* ± 0.02
$VCAM\text{-}1^{-/-}$	6.62 ± 1.01	2.11 ± 0.65	0.05 ± 0.002	0.82 ± 0.15	1.23 ± 0.018	0.92 ± 0.06	0.06 ± 0.002	1.405 ± 0.001
BALf (2 lungs)								
WT	8.5 ± 0.42	1.48 ± 0.05	1.39 ± 0.17	1.14 ± 0.07	0.78 ± 0.04	3.83 ± 0.08	0.04 ± 0.001	0.501 ± 0.001
$\alpha4^{\hat{a}\acute{I}LE/\hat{a}\acute{I}LE}$	0.82* ± 0.17	0.09* ± 0.002	0.04* ± 0.016	0.06* ± 0.001	0.46* ± 0.07	0.019* ± 0.001	0.002* ± 0.0001	0.05* ± 0.001
$B2^{-/-}$	0.72* ± 0.03	0.08* ± 0.002	0.048* ± 0.001	0.141* ± 0.04	0.066* ± 0.001	0	0	0.649 ± 0.05
$VCAM\text{-}1^{-/-}$	8.59 ± 1.67	1.59 ± 0.54	1.46 ± 0.03	0.97 ± 0.21	0.63 ± 0.004	2.54 ± 0.001	0.02 ± 0.001	0.58 ± 0.02

Values represent total numbers (in millions) of different cells and their subsets that migrated from blood (~ 2 mL) to lung interstitium (both lungs) and to BALf (both lungs) of WT (wild-type), $\alpha4^{\hat{a}\acute{I}LE/\hat{a}\acute{I}LE}$, $\beta2^{-/-}$ and $VCAM\text{-}1^{-/-}$ mice. Recruitment of total WBCs (white blood cells) is less in $\alpha4^{\hat{a}\acute{I}LE/\hat{a}\acute{I}LE}$ lung as we as in BALf compared to WT and VCAM-1 mice. In $\beta2^{-/-}$ lungs, except for T cells and eosinophils, all other cells were increased in number

$^{*}p < 0.01$ compared with post-OVA control. Recruited T cells post-OVA (both $CD4^{+}$ and $CD8^{+}$) were CD45RC-negative (memory) in control lung and BALf, but the few T cells in $\alpha4^{\hat{a}\acute{I}LE/\hat{a}\acute{I}LE}$ and $\beta2^{-/-}$ mice were mostly CD45RC-positive (nave). Note the differences between α4- and β2-deficient mice in total LP (lung parenchyma) cell content. Mφ denotes macrophage, $n = 12$/genotype in $\alpha4^{\hat{a}\acute{I}LE/\hat{a}\acute{I}LE}$, $\beta2^{-/-}$, and WT and 7/group in $VCAM\text{-}1^{-/-}$

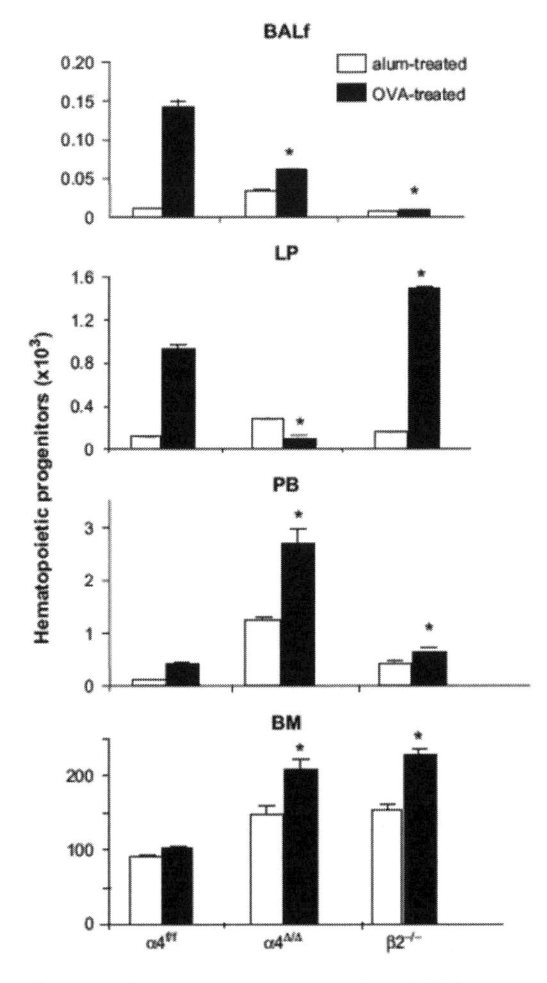

Fig. 3.11 CFU-Cs from alum-treated and OVA-treated mice. Total number of CFU-C per organ was calculated using the number of colonies/number of cells plated, normalized to the total number of cells counted in single cell suspensions obtained from 2 femurs (*BM*), 2 mL of blood (*PB*), lung parenchyma (*LP*), and BALf from both lungs from alum-treated and OVA-treated mice, $^{*}P < 0.05$ compared to OVA-treated control

cytokines and eotaxin in plasma and in BALf in all OVA-treated mice (Fig. 3.12b). After OVA challenge in control mice, a marked increase was seen in Th2 cytokine levels (i.e., IL-5, IL-13, and eotaxin) in plasma and BALf, but very low levels in α4Δ/Δ mice, consistent with absence of effector cell recruitment in BALf.

$\alpha4^{\Delta/\Delta}$ T Cells Fail to Show Proliferative Response to OVA but Not to Non-Th2 Stimuli

To ascertain the intrinsic functional status of α4-deficient T lymphocytes, proliferative responses to specific (OVA) (Fig. 3.12c) and to nonspecific stimuli (anti-CD3/CD28, PMA/ionomycin) (Fig. 3.13) were assayed. Although α4Δ/Δ splenocytes from OVA-sensitized mice proliferate in response to nonspecific stimuli, they failed to proliferate in response to increasing concentrations of OVA, in contrast to control cells, which efficiently proliferate in response to both nonspecific and Th2 stimuli. When Th2 cytokines were measured in the culture supernatant of splenocytes 4 days post-incubation with OVA, levels of all Th2 cytokines remained at baseline levels when α4Δ/Δ cells were used (Fig. 3.12d), in contrast to controls.

Adoptive Transfer of CD4+ Splenocytes from Sensitized α4Δ/Δ Mice Fails to Restore Asthma in Naïve Control Mice upon Their Subsequent Challenge

To test whether the failure of α4Δ/Δ mice to develop asthma is due to an intrinsic inability of α4Δ/Δ splenocytes to respond to OVA and recruit all other cell subsets to lung and BALf, we adoptively transferred (Fig. 3.14) sensitized α4Δ/Δ donor cells (either unpurified α4+ splenocytes or CD4+–purified α4+ splenocytes) from α4f/fMxcre_ (α4+/+) mice to naïve α4+/+ or α4Δ/Δ mice by i.v. injection. While sensitized α4+/+ splenocytes restored asthma phenotype after challenge in both naïve α4+/+ and α4Δ/Δ recipients, sensitized CD4+α4Δ/Δ splenocytes transferred to both naïve α4f/f and α4Δ/Δ recipients failed to cause asthma upon further OVA challenge (Fig. 3.14a, lower line graphs). Analysis of BALf after the adoptive transfer of α4+ cells and OVA challenge revealed an increase in Th2 cytokines (Fig. 3.14b) and an accumulation in BALf of a

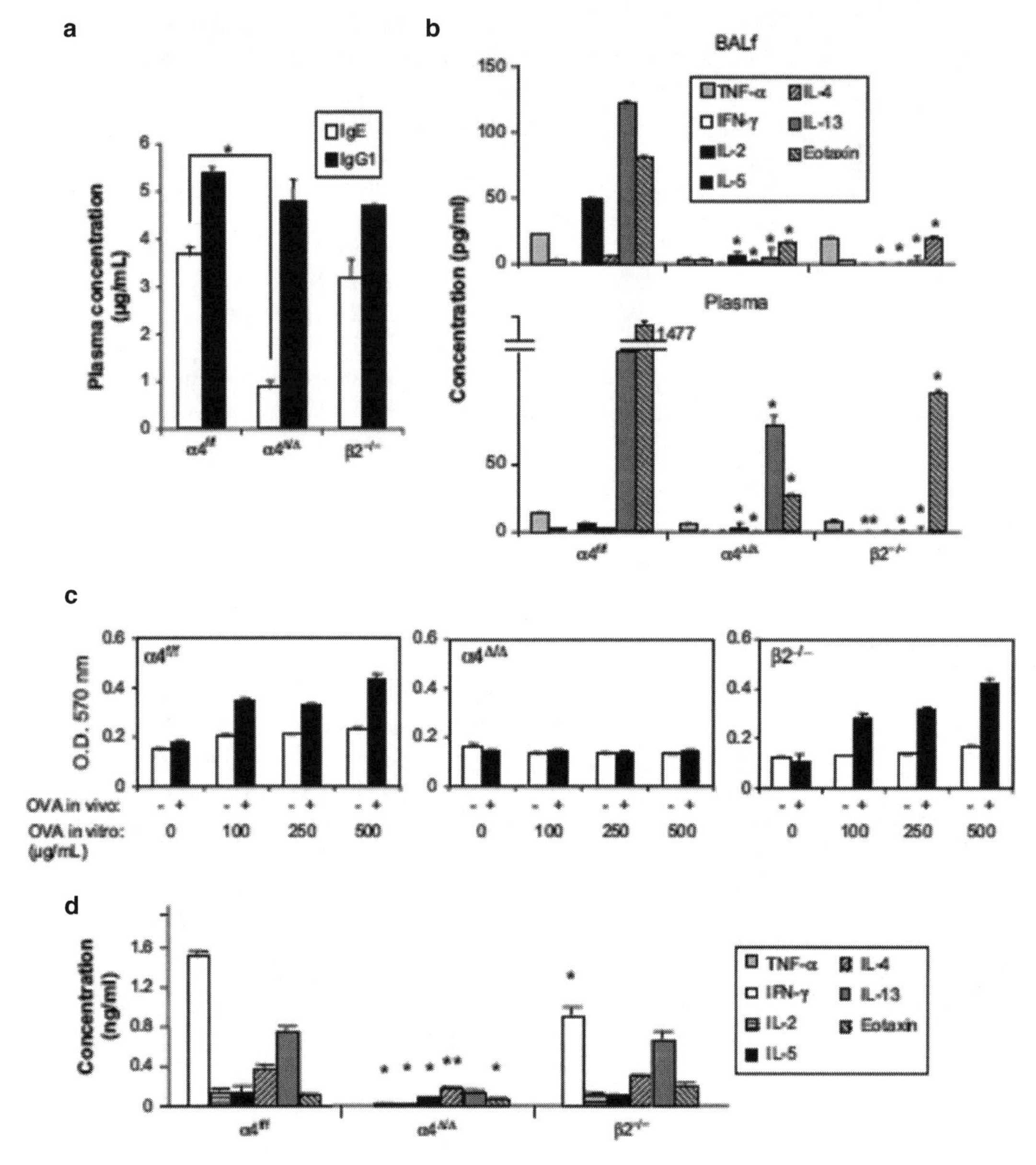

Fig. 3.12 OVA sensitization responses in control ($\alpha 4^{f/f}$) versus $\alpha 4^{\Delta/\Delta}$ and β2-deficient mice. (**a**) Neonatally ablated $\alpha 4^{\Delta/\Delta}$ mice show significantly decreased OVA-specific IgE levels ($^{*}P<0.05$ relative to control) in plasma, but IgG_1 levels were similar in all genotypes, $n=12$. Levels were undetectable in plasma from alum-treated mice in all groups and are not shown. (**b**) Th1 and Th2 cytokines and eotaxin levels in BALf and plasma of OVA-treated mice measured by ELISA and cytometric bead array. Th2 cytokines were lower of $\alpha 4^{\Delta/\Delta}$ and $\beta 2^{-/-}$ mice compared to controls, $^{*}P<0.01$ compared with post-OVA control, $n=12$/genotype. All cytokines in alum-treated mice were below the level of detection. (**c**) Splenocytes from mice of three genotypes with either in vivo saline sensitization or OVA sensitization/challenge were allowed to proliferate in vitro with increasing concentrations of OVA over the course of 4 days. Proliferation was assessed by incorporation of MTT tetrazolium (O.D. 570 nm). Only $\alpha 4^{\Delta/\Delta}$ cells failed to show appreciable proliferative changes, $^{*}P<0.001$ relative to control cells, $n=6$/genotype. (**d**) Cytokines in supernatants from the in vitro OVA-stimulated splenocyte cultures (500 μg/mL OVA) in Fig. 3.12c. Only $\alpha 4^{\Delta/\Delta}$ cells failed to generate increased levels of cytokines, $^{*}P<0.001$ relative to control, $n=6$ mice/genotype

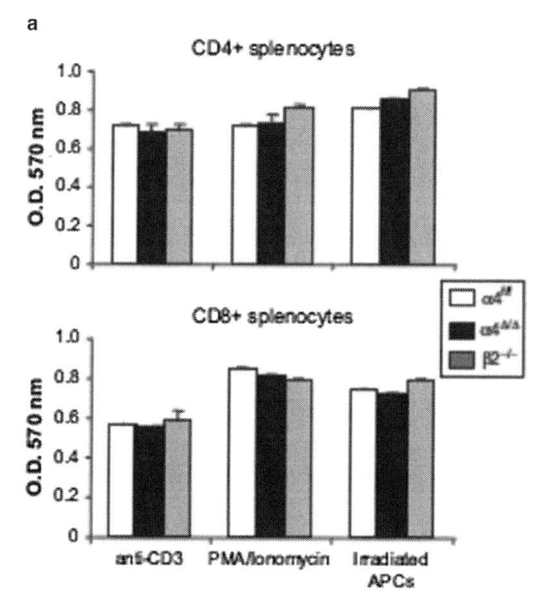

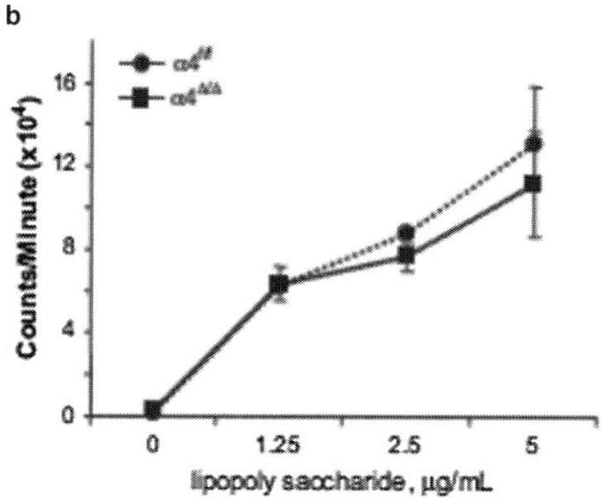

Fig. 3.13 Proliferation responses of splenocytes from control versus knockout mice. $CD4^+$ and $CD8^+$ splenocytes were positively enriched using MACS and activated in vitro with different stimuli: 1 µg/mL anti-CD3/CD 28, 10 ng/mL PMA/1 µM ionomycin, and allogeneic irradiated stimulator splenocytes (APCs). (**a**) shows similar proliferative responses (measured by Cell-Titer96 Assay and OD measurement at 570 nm) in all three genotypes. (**b**) $CD19^+$ cells from control and $\alpha4^{\Delta/\Delta}$ mice enriched by MACS were assayed for proliferative responses to increasing doses of *E. coli* LPS by [^{3}H]-thymidine incorporation. Similar proliferative responses are seen ($n = 5$mice/group)

mononuclear cell population ($4.45 + 0.12 \times 10^6$), predominantly of lymphocytes and a few other cells (macrophages and neutrophils), but no eosinophils.

Lack of Increased VCAM-1 Expression in Response to OVA in α4Δ/Δ Lung

Upregulated VCAM-1 expression in lung vasculature in response to OVA is another hallmark of allergic asthma [18, 45], believed to be implicated in the enhancement of inflammatory cell interactions with the endothelium and the preferential recruitment of memory cells or mast cells [63]. Additionally, soluble VCAM-1 (sVCAM-1) may be important in promoting the allergic late-phase eosinophilia in a positive feedback loop of interaction between α4 integrin and its ligand VCAM-1 [64]. We documented that VCAM-1 expression in α4Δ/Δ lung, as well as sVCAM-1 in BALf and plasma of α4Δ/Δ mice after OVA challenge, was significantly reduced, compared to that found in control samples (Fig. 3.15).

Differences Between Responses of α4Δ/Δ, CD18–/–, or VCAM-1D/D Mice to Acute OVA Challenge

In addition to control (α4f/f), CD18–/– (CD18 KO) mice [43, 65] were used for comparison, to explore whether these two integrins have overlapping or redundant roles in inflammatory recruitment of leukocytes in acute asthma. Like the other groups, CD18–/– mice responded systemically by increasing nucleated cells and progenitors in their BM and PB (Figs. 3.10a and 3.11). However, as previously described [43], CD18–/– mice failed to develop AHR and to accumulate eosinophils and lymphocytes in BALf (Fig. 3.10a). In contrast to BALf, we found significant recruitment of both mature myeloid cells (Table 3.1) and progenitors in the lung of CD18–/– mice post-OVA (Figs. 3.10a and 3.11). Nevertheless, the proportion of recruited cells in the lung was 8 % of circulating cells, in contrast to 29 % in control mice, suggesting that the recruitment of CD18–/– cells in allergic lungs, despite the significant numbers, was not as efficient as that seen with normal cells. In addition to the levels of cytokines (in response to acute OVA challenge) obtained from ELISA, we also measured Th1 versus Th2 cell numbers in α4Δ/Δ and CD18–/– mice to exclude the

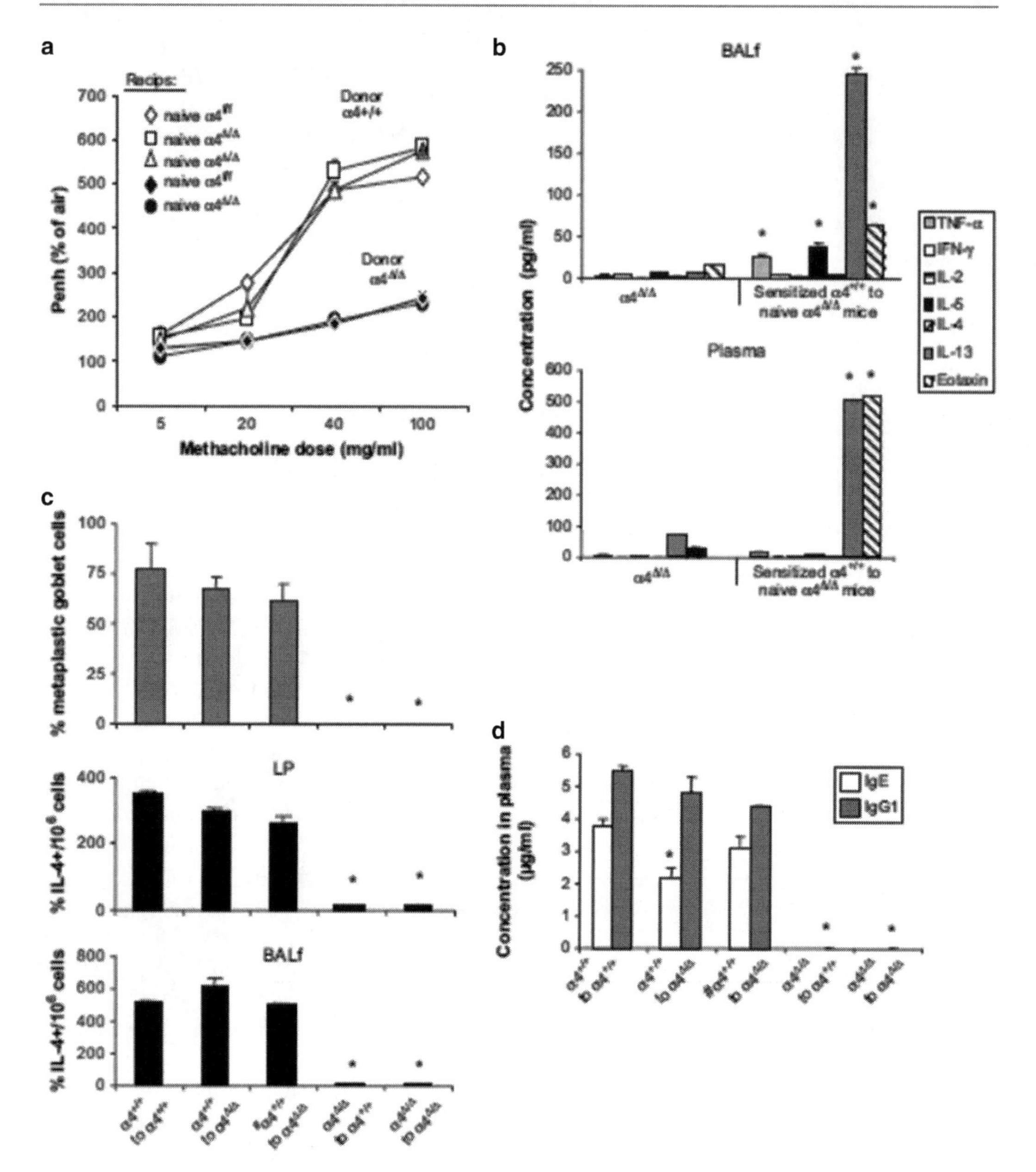

Fig. 3.14 Adoptive transfer of sensitized CD4$^+$ α4$^+$ splenocytes in naïve α4$^{\Delta/\Delta}$ mice stores asthma response. (**a**) Naïve α4$^{\Delta/\Delta}$ recipients receiving CD4$^+$ splenocytes (□) or unpurified splenocytes (Δ) from sensitized α4$^{+/+}$ donors showed increased AHR after challenge. By contrast, when sensitized α4$^{\Delta/\Delta}$ donor cells were given to naïve recipient mice, neither control α4$^{\Delta/\Delta}$ (♦) nor α4$^{\Delta/\Delta}$ (•) showed increased AHR after challenge. Naïve α4$^{f/f}$ recipients given sensitized CD4$^+$ α4$^{+/+}$ splenocytes (◇) were considered positive controls. (**b**) Th1 and Th2 cytokines and eotaxin levels from plasma of BALf from naïve α4$^{\Delta/\Delta}$ mice after the adoptive transfer of sensitized CD4$^+$ α4$^+$ splenocytes are shown compared to nontransplanted OVA-treated α4$^{\Delta/\Delta}$ mice, $^*P < 0.001$ relative to α4$^{\Delta/\Delta}$ without adoptive transfer (**c**) *Upper panel*: development of Alcian blue-stained mucus-laden metaplastic goblet cells in lung airway lumina of unsensitized mice given cells from sensitized α4$^{+/+}$ donors (**a**, *upper line graphs*). *Middle* and *lower panels*: percentage of IL4+ cells/10^6 cells (ELISPOT assay) show a similar trend in LP and BALf, $n = 5$/group. $^*P < 0.05$ comparing naïve recipients given cells from sensitized Δ/Δ donors to recipients given cells from sensitized +/+ donors. (**d**) OVA-specific IgE and IgG1 in plasma were increased in all recipients of α4$^{+/+}$ donor cells, but not in recipients receiving α4$^{\Delta/\Delta}$ cells (#purified CD4$^+$ cells, others are unpurified). For **a–d**: $n = 5$ mice/group

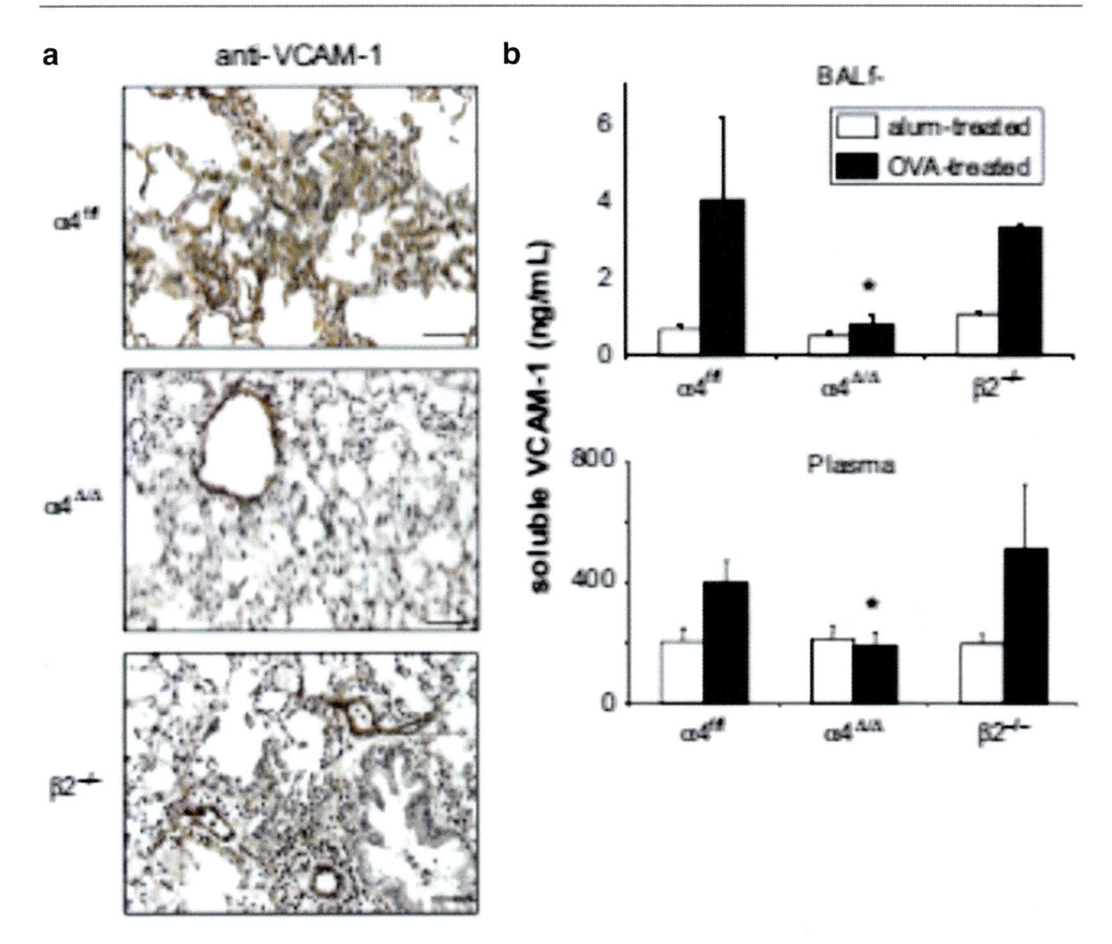

Fig. 3.15 VCAM-1 expression in lung sections and sVCAM-1 levels in BALf and plasma in response to OVA. (**a**) Upregulation of VCAM-1 is seen in control and $\beta^{2-/-}$ but not in $\alpha 4^{\Delta/\Delta}$ prefixed, frozen lung tissue sections. (**b**) Soluble VCAM-1 levels in BALf and plasma show comparable increases in control and $\beta 4^{-/-}$ mice but remain unchanged in $\alpha 4^{\Delta/\Delta}$ mice. $^{*}P < 0.001$ compared to OVA-treated control, $n = 6$ mice/group

possibility that low levels of IL-4 in BALf may be due to bound IL-4 in the lung interstitium. IL-4+ (Th2) and IFN-γ+ (Th1) cells in BALf and the lung quantitated by ELISPOT (Fig. 3.10b) were again decreased in both α4Δ/Δ and CD18–/– mice. TNF-α was also reduced in BALf of α4Δ/Δ compared to controls, whereas Th2 cytokines in plasma were increased only in OVA-treated control mice, but not in similarly treated α4Δ/Δ and CD18–/– mice (Fig. 3.12b). To clarify the role of VCAM-1 upregulation in asthma, we induced and similarly evaluated asthma phenotype in VCAM-1-deficient mice [66]. In contrast to α4-deficient mice, all the cellular and humoral hallmarks of asthma were present in these mice (Fig. 3.16).

Discussion

Because prior data from studies using antibodies to α4 integrins provided inconsistent results varying with the type, dose, and route of administration of antibody used [44–51], in the present study we explored the role of α4 integrins in the development of acute allergic asthma using a conditional α4-integrin knockout mouse. Further, parallel studies in CD18–/– and VCAM-1-deficient mice provided novel mechanistic insights about the selective roles of the two classes of integrins in allergen sensitization and inflammatory cell recruitment to allergic lung and the role of VCAM-1 in asthma.

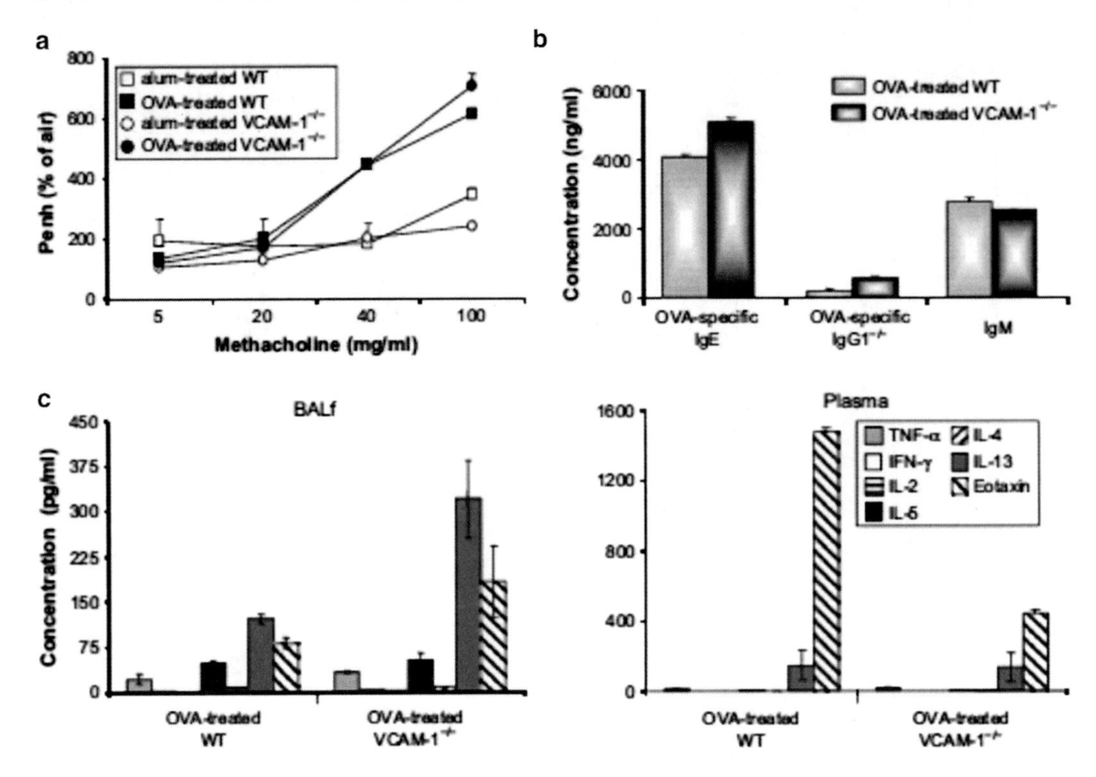

Fig. 3.16 Acute asthma induction in VCAM-1-deficient mice compared to OVA-treated controls. (**a**) Response to methacholine was determined by noninvasive plethysmography. (**b**) Plasma levels of IgE, IgG1, and IgM in VCAM-1$^{-/-}$ mice and their controls after OVA sensitization. (**c**) The concentrations of Th1 and Th2 cytokines in BALf and plasma were determined by ELISA (as in Fig. 3.12). The number of mice studied was 7/genotype

Studies addressing OVA sensitization were of interest in α4Δ/Δ mice. Although an adequate systemic cellular response was seen in BM and PB (Fig. 3.10a), OVA-specific IgE in plasma (Fig. 3.12a) was only modestly increased. Normal levels of OVA-specific IgG1 (but low IgE) in α4Δ/Δ mice support differential regulation (i.e., differential sensitivity to IL-4) of IgG1 versus IgE production noted previously [67–69]. It is of interest that although both IgE and IgG1 are IL-4 dependent, IL-4/IL-13 double-deficient mice and IL-4/R α4−/− mice produced IgE and IgG1 responses [70], suggesting IL-4-/IL-3-independent responses. Therefore, adequate production of IgG1 and decreased IgE are neither incompatible with low levels of IL-4 nor do they predict on their own the asthma outcome. For the systemic BM response, we speculate that increased sensitivity of α4Δ/Δ cells to tested or untested cytokines (e.g., IL-3, IL-9) and/or decreased levels of inhibitory cytokines may have been responsible, whereas the modest IgE levels seen in these mice could reflect an inefficient sensitization process largely derived from regional lymph node cells [62]. In addition to reduced IgE, low levels of Th2 cytokines were seen in plasma samples from OVA-treated α4Δ/Δ (Fig. 3.12b) mice. As the initiation of asthmatic reaction is Th2 dependent, it would seem that α4-deficient mice are unable to sensitize a sufficient number of Th2 cells to raise PB cytokine levels similar to those seen in controls. CD4+ and CD8+ splenocytes from α4Δ/Δ mice after primary in vivo OVA sensitization were unable to proliferate in response to OVA in vitro (Fig. 3.12c), although their proliferation to nonspecific stimuli, like anti-CD3/CD28, PMA/ionomycin, or allogeneic splenocytes from BALB/c mice(APCs), remained similar to that of control cells (Fig. 3.13a).

The response of α4Δ/Δ CD19+ splenocytes to LPS in vitro was also normal (Fig. 3.13b). In contrast to α4Δ/Δ, CD18−/− splenocytes showed similar proliferative responses to both specific (OVA) and nonspecific stimuli (anti-CD3, PMA, etc.) compared to controls [60]. Likewise, Th1 and Th2 cytokines and eotaxin were increased in culture supernatants of both control and CD18−/− cells, but not in culture supernatants from α4Δ/Δ splenocytes (Fig. 3.12d).

Nevertheless, in CD18−/− mice, circulating Th2 cytokines were surprisingly low for unclear reasons. We speculate that an increased uptake by tissue-wide distribution of inflammatory cells at steady state in these mice may be responsible. Migration impairments noted in α4Δ/Δ mice could be attributed to either intrinsic migratory defects of all α4Δ/Δ leukocyte subsets (lymphocytes, monocytes, eosinophils), or migratory deficiency of only one population upstream (i.e., lymphocytes), that normally triggers a chemotactic response for other populations downstream in a sequential fashion. The data from adoptive transfer of both α4+/+ and α4Δ/Δ cells (Fig. 3.14) provide further insight into the OVA-specific unresponsive state of CD4+ α4Δ/Δ cells and collectively indicate (1) the APCs in α4Δ/Δ mice were able to reactivate transferred α4+/+ CD4+ cells for asthma induction, (2) eosinophil migration to airway lumen even in the presence of chemotactic stimuli is dependent primarily on α4 integrin, as no eosinophils migrated with transferred α4+/+ CD4+ cells, (3) eosinophils are not necessary for the development of the acute asthma phenotype consistent with recent data [59, 60], and (4) the lack of asthma phenotype development in α4Δ/Δ mice is not contributed by defective responses from airway resident cells including alveolar macrophages in this model, as previously suggested [48, 71]. Thus, the failure of asthma development upon transfer of sensitized α4Δ/Δ cells suggests intrinsic defects in CD4+ responses and putative inefficiency in T- and B-cell interactions. Lack of OVA-specific proliferation in vitro by sensitized splenocytes from α4Δ/Δ mice likely reflects inefficient responses from α4-deficient CD4+ T cells. Consistent with these data are previous studies suggesting VLA-4-dependent costimulatory effects on CD4+ T-cell activation [72–75]. Recent independent studies also show evidence of an important role of VLA-4 in the generation of the immune response through its involvement in the formation of immune synapse [76], where α4b1 is relocated to the peripheral supramolecular activation complex (pSMAC) during immune synapse formation between APCs and T lymphocytes. An inhibition of asthma similar in many aspects (i.e., cytokine responses, IgE responses) to that seen in our α4Δ/Δ mice was recently described after engagement (by agonistic Ab) of CD137 (4-1BB), an inducible T-cell costimulatory receptor [77].

In this case, a more complete anergy of OVA-specific CD4+ T cells was induced than the one seen in α4Δ/Δ mice. Impairment in IL-4 and/or TNF-α secretion or other cytokines by α4Δ/Δ cells in response to OVA may also be responsible for lack of VCAM-1 upregulation in α4Δ/Δ lung post-OVA (Fig. 3.15), thereby preventing further cell sequestration and cytokine responses [78]. In addition to lung endothelial cells, VCAM-1 is upregulated in bronchial epithelial cells through their interaction with eosinophils in normal mice [79]. Lack of α4/VCAM-1 interaction, likely due to a negative feedback loop in the α4Δ/Δ mice, was considered significant in the inhibition of development of the Th2 effector arm critical for asthma onset and maintenance, and α4 deficiency may prevent cross talk between the inflammatory leukocyte and the endothelium important for acute asthma development.

Whether additional factors, i.e., an increase in ADAM8, a sheddase for VCAM-1, are responsible for increase in sVCAM-1 in plasma [80] in other mice except α4Δ/Δ is also unclear. Nevertheless, asthma was not inhibited in our VCAM-1-deficient mice, suggesting that VCAM-1 upregulation in normal mice is likely a secondary event with no causative involvement in asthma. Of interest, anti-VCAM-1 treatment did not block IL-4-induced eosinophil accumulation in rat pleural cavities, in contrast to anti-rat α4 mAb. These and our data suggest dissociation between VCAM-1 upregulation and eosinophil accumulation, which is dependent on α4 integrin. In the present study, both α4Δ/Δ and CD18−/− mice failed to develop a typical asthma phenotype in response to acute OVA challenge (Figs. 3.9 and 3.10). Upon dissecting the various criteria jointly leading to the composite asthma phenotype, we found that the migratory behavior of different cell subsets in the two types of deficient mice was quite distinct. In fact, the pattern of migratory responses from blood to lung and to airway lumen displayed by α4Δ/Δ mice differed from both control and CD18−/− mice. In contrast to control and in contrast to Ab studies [49], all α4Δ/Δ leukocyte subsets (T cells, B cells, monocytes, neutrophils, eosinophils) failed to migrate from circulation to both lung and airway lumen, whereas in CD18−/− mice, only T cells and eosinophils failed to migrate to lung and lumen. For other CD18−/− cells (B cells, monocytes, neutrophils, and macrophages), α4 integrins mediate all of the CD18-independent migration to lung [49], but there is interstitial migratory failure from lung to airway lumen (Table 3.1). In summary, our data demonstrate that α4 integrins are necessary for all cell subset migration to allergic lung, and their absence attenuates the sensitization process and ancillary functions of sensitized cells. Further, our data clarify the subset-specific integrin usage for recruitment of cells from blood to lung interstitium and then to airway lumen. In agreement with prior data [59, 81, 82], eosinophil migration to BAL fluid is not essential for AHR and is an α4- but not a VCAM-1-dependent process.

Conclusions

The lack of α4 integrin not only impedes the migration of all white cell subsets to lung and airways but also prevents upregulation of vascular cell adhesion molecule-1 (VCAM-1) in inflamed lung vasculature and, unlike β2, attenuates optimal sensitization and ovalbumin-specific IgE production in vivo. As VCAM-1 deficiency did not protect mice from asthma, interactions of α4b1+ or α4b7+ cells with other ligands are suggested.

Subchapter 2: Role of E-, L-, and P-Selectins in the Onset, Maintenance, and Development of Acute Allergic Asthma

Summary of the Study

Objective: The recruitment of leukocytes from circulation to sites of inflammation requires several families of adhesion molecules among which are selectins expressed on a variety of cells. In addition, they have also been shown to play key roles in the activation of cells in inflammation. *Methods*: To explore the collective role of E-, L-, and P- selectins in OVA-induced Th2-mediated response in acute asthma pathophysiology, ELP−/− mice were used and compared with age-matched wild type (WT). *Results:* Asthma phenotype was assessed by measuring pulmonary function, inflammation, and OVA-specific serum IgE, which were completely abrogated in ELP−/− mice. Adoptive transfer of sensitized L-selectin + CD4+ T cells into naïve ELP−/− mice which post-OVA challenge, developed asthma, suggest that L-selectin may be critically involved in the onset of Th2 response in asthma. Tissue resident ELP-deficient cells were otherwise functionally competent as proved by normal proliferative response. *Conclusions:* Comparative studies between ELP−/− and WT mice uncovered functional roles of these three integrins in inflammatory response in allergic asthma. All three selectins seem to impede inflammatory migration, while only L-selectin also possibly regulates the activation of specific T-cell subsets in the lung and airways.

Background and Scope of the Study

The respiratory drug market is dominated by asthma and its exacerbations (worsening symptoms, rescue medication use, and emergency department visits or hospitalizations). Asthma is the third leading cause of death in both developed and developing countries, and annual direct and indirect cost of healthcare is more than $50 billion in the USA alone. A small percent of non-responders (10 %) account for greater than 50 % of healthcare costs, and it is for these and other patients with exacerbations that alternative target redressal is not just necessary but indispensible given the healthcare costs [83].

In particular, there is a need to develop drugs that control the underlying inflammatory and destructive processes. Rational treatment depends on understanding the underlying disease process, and there have been recent advances in understanding the cellular and molecular mechanisms that may be involved to look for better drug targets [84]. Inflammation is key to etiology of most respiratory disorders, and there is a fine balance between the beneficial effects of inflammation cascades and inflammation cascades lead to the development of diseases such as chronic asthma, rheumatoid arthritis, psoriasis, multiple sclerosis, and inflammatory bowel disease. The specific characteristics of inflammatory response in each disease and site of inflammation may differ, but recruitment and activation of inflammatory cells and changes in structural cells remain a universal feature along with a concomitant increase in the expression of components of inflammatory cascade including cytokines, chemokines, growth factors, enzymes, receptors, adhesion molecules, and other biochemical mediators [85]. The chronic airway inflammation of asthma is unique in that the airway wall is infiltrated by T lymphocytes of the T-helper (Th) type 2 phenotype, eosinophils, macrophages/monocytes, and mast cells. Accumulation of inflammatory cells in the lung and airways, epithelial desquamation, goblet cell hyperplasia, mucus hypersecretion, and thickening of submucosa resulting in bronchoconstriction and airway hyperresponsiveness are important features of asthma [86]. Both cells from among the circulating leukocytes such as Th2 lymphocytes, mature plasma cells expressing IgE, eosinophils [87], and neutrophils as well as local resident and structural cells constituting the "respiratory membrane" (airway epithelial cells, fibroblasts, resident macrophages, bronchial smooth muscle cells, mast cells, etc.) contribute to the pathogenesis of asthma [88]. Cross-linking of IgE receptors on mast cells releases histamines, prostaglandins, thromboxane, and leukotrienes, leading to bronchoconstriction, vasodilation, and mucus secretion.

A cascade of interactions between cells and soluble molecules result in bronchial mucosal inflammation and lead to airway hyperresponsiveness [31]. Leukocyte emigration into lung is an important event in the pathogenesis of asthma, likely mediated by a series of leukocyte adhesion molecule interactions with endothelium of which the various ICAMs, adhesion molecules [89], and selectins [90, 91] have been found to be critically important. Among the integrins, $\alpha4$ is key in initial signaling for sensitization as well as migration for the onset and development of a full-blown acute asthma phenotype as well as airway remodeling in chronic asthma, while $\beta2$ integrins are solely required for mechanical migration of leukocytes [92, 93].

During inflammatory recruitment in the lung and airways, the initial contact of leukocytes with the endothelium is mediated by selectins and their ligands inducing the rolling of leukocytes along the vessel wall [46, 57, 58, 94, 95]. This rolling phenomenon is a prerequisite for the subsequent firm adhesion and transmigration, which is mediated by members of the integrin family, e.g., $\beta2$ integrins, and immunoglobulin gene superfamily, e.g., intercellular adhesion molecule-1 (ICAM-1) [96–98]. Peribronchial inflammation contributes to the pathophysiology of allergic asthma. In many vascular beds, adhesive interactions between leukocytes and the endothelial surface initiate the recruitment of circulating cells. Such movement is believed to follow a coordinated and sequential molecular cascade initiated, in part, by the three members of the selectin family of carbohydrate-binding proteins: E-selectin (CD62E), L-selectin (CD62L), and P-selectin (CD62P). The role of selectins in neutrophil trafficking in the lungs was frequently considered negligible since the narrow pulmonary capillaries cannot accommodate the typical selectin-mediated rolling phenomenon. Selectins especially did not seem necessary in lung neutrophil sequestration since the deceleration of circulating neutrophils prior to their firm adherence was effectively achieved by their mechanical retention [91]. Yet a large body of experimental data demonstrating that selectin inhibition (via the use of blocking antibodies or selectin antagonists or transgenic knockout of one or more selectins) frequently protected animals from acute lung injury [Subset-Specific Reductions in Lung Lymphocyte Accumulation Following intratracheal antigen Challenge in Endothelial Selectin-Deficient Mice] [95, 99, 100]. P-selectin-mediated platelet-neutrophil interactions are critical to the development of ALI, to which they contribute by enhancing their respective activations, which trigger the production of TXA2 and other inflammatory mediators [101].

In addition to being responsible for the rolling phenomenon, selectins have been implicated as important signaling molecules involved in leukocyte activation as well. The selectin gene family closely linked on mouse chr1 encodes three structurally related proteins that display differential spatial and temporal expression within the vascular system. Endothelial cells express E-selectin, platelets express P-selectins, and leukocytes express L-selectin. Whereas P-selectin (CD62P) is rapidly mobilized to the surface of activated endothelium or platelets, E-selectin (CD62E) expression is induced by inflammatory cytokines. L-selectin (CD62L) is constitutively expressed on most leukocytes. Selectin expression can be upregulated under inflammatory conditions in experimental animals [102–105] and humans [106]. Although there appear to be some variations in E-selectin expression depending on the disease state, the general consensus from these studies is that E-selectin is most strongly expressed on endothelial cells of portal tract vessels and hepatic venules and to a lesser degree on sinusoidal lining cells [102]. In contrast, L-selectin is constitutively present on most types of leukocytes [107]. In vitro studies showed that E-selectin on endothelial cells can induce upregulation of Mac-1 (CD11b/CD18) on neutrophils [108]. In addition, L-selectin ligation or cross-linking induced upregulation of Mac-1 and priming for superoxide formation [97]. Thus, E-, P-, and L-selectins may have an important function in Th2 response in OVA-induced asthma. The objective of this study was to investigate the functional significance of E-, P-, and L-selectins in an experimental model of OVA-induced lung injury. OVA induces a Th2-mediated inflammatory response characterized by inflammatory recruitment in the lungs and airways, mucus over-secretion, and airways hyperresponsiveness. Our previous studies [92, 93] have shown the involvement of various families of adhesion molecules, viz., $\alpha 4\beta 1$, $\beta 2$, and VCAM-1, which facilitate leukocyte transmigration, adherence to parenchymal cells, and Th2 response, in the pathophysiology of various inflammatory disease models such as allergic asthma and aseptic peritonitis. This study elucidates the effect of deletion of all three selectins in the development of allergic asthma phenotype in mice.

Results in a Nutshell

Asthma phenotype was assessed by measuring pulmonary function, inflammation, and OVA-specific serum IgE, which were completely abrogated in ELP−/− mice. Adoptive transfer of sensitized L-selectin + CD4+ T cells into naïve ELP−/− mice which post-OVA challenge, developed asthma, suggest that L-selectin may be critically involved in the onset of Th2 response in asthma. Tissue resident ELP-deficient cells were otherwise functionally competent as proved by normal proliferative response.

Results

Complete Abrogation of Composite Asthma Phenotype in KO Mice

Our first aim was to find out whether composite asthma phenotype developed (Fig. 3.17) in the absence of the triple selectins in a knockout mouse which should indicate what role they have in the onset of the disease pathophysiology. We found that following allergen challenge, compared to the WT mice which develop all the hallmarks of the composite asthma phenotype, the yardsticks quantified in the ELP−/− mice were all similar to the saline-treated baseline controls (Fig. 3.18a–c), viz., there was no significant

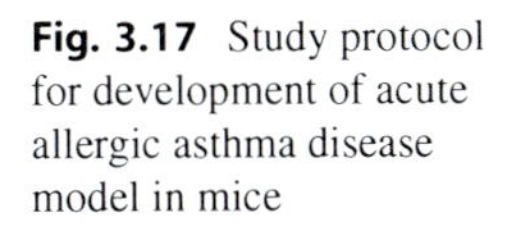

Fig. 3.17 Study protocol for development of acute allergic asthma disease model in mice

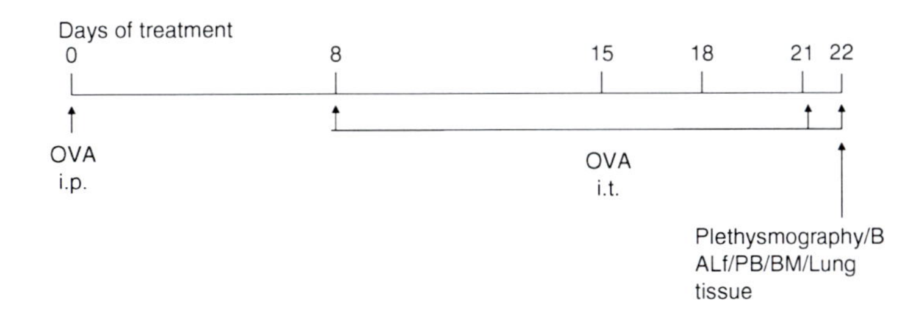

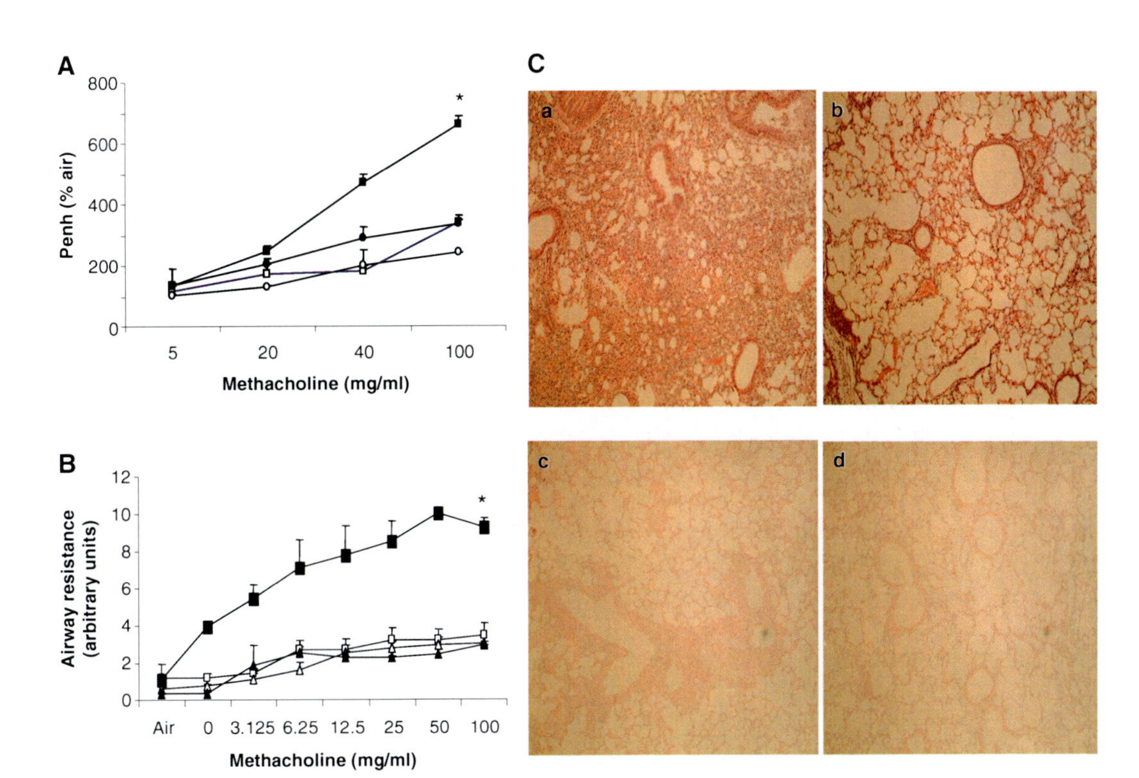

Fig. 3.18 Non-development of composite asthma phenotype in ELP−/− mice. (**a**). Pulmonary function test by noninvasive whole body plethysmography. Functional response to increasing doses of methacholine show no increase in response from baseline in ELP−/− mice in comparison with WT post-OVA. WT + alum (*white square*), WT + OVA (*black square*), ELP−/− + alum (*white square*), and ELP−/− + OVA (*black square*); (**b**). *Pulmonary function test by invasive plethysmography.* Functional response to increasing doses of methacholine show no increase in response from baseline in ELP−/− mice in comparison with WT post-OVA [WA = WT with alum and saline treatment, EA = ELP−/− mice same placebo treatment; WO = WT sensitized and challenged with OVA, EO = OVA treatment with ELP−/− mice] WA (*white square*), WO (*black square*), EA (*white triangle*), EO (*black triangle*); (**c**). *Lung histology. Panel a, b* H&E staining of paraffin lung sections a WT + OVA, b ELP−/− + OVA; *Panel c, d* Alcian blue staining of paraffin lung sections. C. WT + OVA, d ELP−/− + OVA

increase in either Penh or RL in response to increasing doses of aerosolized methacholine as measured by a plethysmometer nor inflammation in the airways or lung parenchyma as detected H&E staining of histological sections of the lung post-OVA in the ELP−/− mice versus WT.

Once it was clear that the preclinical parameters for acute allergic asthma failed to develop in the knockout mouse, we embarked on dissecting the patterns of inflammatory recruitment from bone marrow to peripheral blood to lung parenchyma to the lung interstitium which is the stage for the drama of the disease pathology to manifest itself. Tables 3.1, 3.2, 3.3 that despite increased number of mature leukocytes and progenitors in BM and circulation

Table 3.2 Total number of cells ($\times 10^6$) in various tissues

	BM/femur	PB/ml	Spleen	BALf/2lungs	LP/2lungs
WT + alum	14.28 ± 2.38	4.17 ± 1.01	103.14 ± 31.95	0.68 ± 0.03	1.8 ± 0.63
WT + OVA	31.35 ± 11.95	7.36 ± 2.01	167.45 ± 45.97	8.4 ± 2.22	3.14 ± 0.95
$ELP^{-}/^{-}$ + alum	29.54 ± 2.08*	15.31 ± 0.41*	290.43 ± 38.57*	0.78 ± 0.005	3.79 ± 0.07*
$ELP^{-}/^{-}$ + OVA	33.18 ± 1.06*	15.5 ± 0.77*	247.63 ± 10.36*	0.81 ± 0.013*	5.6 ± 0.17*

Total number of cells ($\times 10^6$) in different tissues
*denotes p value < 0.01 and is considered statistically significant

Table 3.3 Total number of different types of leukocytes ($\times 10^6$) in various tissues

BM/femur	**Mononuclear**	**PMN**	**Eos**			
WT + alum	8.16 ± 1.92	5.13 ± 1.56	0.56 ± 0.02			
WT + OVA	14.68 ± 3.79	11.48 ± 3.3	5.17 ± 1.22			
$ELP^{-}/^{-}$ + alum	19.27 ± 2.48*	9.53 ± 1.02*	0.73 ± 0.06			
$ELP^{-}/^{-}$ + OVA	19.56 ± 5.59∂	13.3 ± 4.11∂	0.31 ± 0.04∂			
PB/ml	**Lymphocyte**	**Monocyte**	**Basophil**	**PMN**	**Eosinophil**	
WT + alum	2.68 ± 0.16	0.14 ± 0.06	0.03 ± 0.009	1.17 ± 0.28	0.12 ± 0.01	
WT + OVA	2.39 ± 0.43	1.6 ± 0.13	0.11 ± 0.003	1.34 ± 0.36	1.9 ± 0.25	
$ELP^{-}/^{-}$ + alum	9.67 ± 2.01*	0.57 ± 0.03*	0.07 ± 0.004	2.41 ± 0.56*	0.17 ± 1.01	
$ELP^{-}/^{-}$ + OVA	9.21 ± 1.09∂	0.49 ± 0.12∂	0.08 ± 0.002∂	5.45 ± 0.38∂	0.25 ± 0.07∂	
BALf/2lungs	**Lymphocyte**	**Monocyte**	**Macrophage**	**PMN**	**Eosinophil**	**Mast cell**
WT + alum	Entirely epithelial cells and monocytes/macrophages					
WT + OVA	1.93 ± 0.38	1.17 ± 0.19	2.11 ± 0.12	1.61 ± 0.33	0.03 ± 0.003	0.003 ± 0.001
$ELP^{-}/^{-}$ + alum	Only epithelial cells and monocytes/macrophages					
$ELP^{-}/^{-}$ + OVA	Only epithelial cells and monocytes/macrophages					
LP/2lungs	**Lymphocyte**	**Monocyte**	**Macrophage**	**PMN**	**Eosinophil**	**Mast cell**
WT + alum	0.2 ± 0.09	0.8 ± 0.02	0.48 ± 0.32	0.3 ± 0.09	0	0
WT + OVA	0.72 ± 0.04	0.44 ± 0.11	0.79 ± 0.11	0.6 ± 0.03	0.57 ± 0.13	0.012 ± 0.001
$ELP^{-}/^{-}$ + alum	0.41 ± 0.02*	1.65 ± 0.35*	1.07 ± 0.04*	0.65 ± 0.02	0	0
$ELP^{-}/^{-}$ + OVA	0.51 ± 0.03	2.01 ± 0.41∂	1.22 ± 0.37∂	01.35 ± 0.03∂	0	0

Number of cells ($\times 10^6$) in various cell subsets in different tissues
*denotes p value < 0.01 and is considered statistically significant

of post-OVA ELP−/− mice, their number was insignificant especially in the airways. Lungs of post-OVA ELP−/− mice show elevated number of monocytes, macrophages, and neutrophils (myeloid population) but no eosinophils and mast cells as opposed to the higher number of these cells in OVA-treated WT (Tables 3.2, 3.3, and 3.4).

ELP−/− B Cells Are Incapable of Sequestering Either Total or OVA-Specific Igs

Since overall allergy depends on cross talk between the T and B cells and a concerted and sequential interplay between the two cell populations, our next question was whether the knockout B cells themselves were otherwise functionally competent. B-cell function was tested by assaying total as well as OVA-specific Igs post-OVA. Figure 3.19 shows that other than total IgG3, all other Igs, notably IgE and IgG1, were significantly inhibited and even IgG3 was decreased compared to WT + OVA values.

Lack of Inflammatory and Th2 Response in KO Mice

Following the analyses of the cells taking part in the development of the allergic inflammation

Table 3.4 Total number of CFU-c in various tissues

	BM/femur	PB/ml	BALf/2 lungs	LP/2 lungs
WT + alum	45,892 ± 1,541	160 ± 34	3 ± 0.2	59 ± 12
WT + OVA	76,894 ± 1,376	1,020 ± 96	872 ± 37	850 ± 14
$ELP^{-}/^{-}$ + alum	61,268 ± 3,497[*]	740 ± 45[*]	11 ± 5[*]	121 ± 47[*]
$ELP^{-}/^{-}$ + OVA	63,550 ± 3,492	845 ± 117[∂]	14 ± 6.3[∂]	269 ± 68[∂]

Number of progenitor cells in various tissues evaluated by CFU-c. The numbers denoting progenitors have been rounded to the nearest whole number. $n = 5$/group
[*]denotes p value < 0.05 compared to WT + alum
[∂]denotes p value < 0.05 compared to WT + OVA

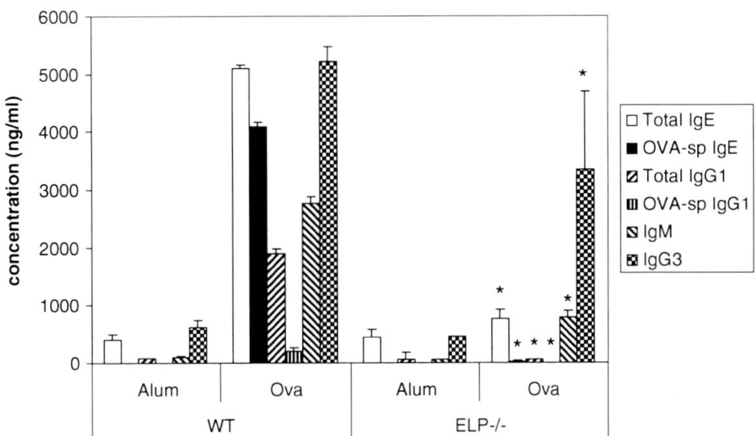

Fig. 3.19 Serum concentration of immunoglobulins in ELP−/− versus WT post-OVA. Serum from infra-orbital bleeding of the experimental mice, collected in heparinized tubes, were frozen at −70°C and later methods described in [41] were used to detect total and OVA-specific immunoglobulin concentration. Presented values are data pooled from two independent experiments with five mice per experimental group measured in dulicate in 96-well format.[*]denotes p value < 0.05 compared to WT + OVA values

itself, both locally and systemically, the next question to be addressed was the actual amount of the cytokine mediators that were being synthesized and released by the inflammatory cells both for propelling their recruitment in the relevant sites and their specific functional manifestation therein. Release of Th2-specific cytokines was significantly downregulated in KO BALf and plasma (Table 3.4) compared to OVA-treated control. Detection of virtually no IL-4-secreting cells in all except LNX (BALf, LP, spleen, BM, MLN, CLN, LNI) and very low number of IFN-γ-secreting cells in all but the peripheral lymph nodes (CLN, LNX, and LNI) was found (Table 3.5 Figs. 3.20 and 3.21).

ELP−/− T Cells Retain Normal Proliferative Response to Mitogenic Stimuli

If the cytokine release by the knockout Th2 cells were unsatisfactory as we found, the next query to be addressed was, "were they functionally competent otherwise?" To assess the functional status of KO T cells, CD4+ T cells were isolated from spleen and peripheral lymph nodes and subjected to proliferation assays to various stimuli (PMA, ionomycin, and anti-CD3/CD28) which showed efficient proliferative response similar to WT (Fig. 3.22a–c). In addition, CD3 level on the T cells in WT and ELP−/− cells before and after OVA is similar (Table 3.5) and no difference in T cytokine production by ELISPOT was observed (data not presented).

Table 3.5 Cytokine concentration (pg/ml) in plasma and BALf post-OVA in WT versus ELP−/−

	Mean	
	WT	ELP−/−
Plasma		
IL-2	41.6 ± 7	4.95 ± 1.75*
IFN-γ	40.89 ± 1.19	2.86 ± 0.88*
TNF-α	9.1 ± 0.4	nd
IL-10	78.05 ± 22.65	6.55 ± 3.05*
IL-4	81.95 ± 0.85	nd
IL-5	244.8 ± 22.8	87.7 ± 6.6*
IL-13	136.1 ± 0.8	nd
Eotaxin	337.33 ± 40.32	nd
MMP-9	96696.45 ± 158.11	70603 ± 40399.7*
SDF-1α	665.6 ± 73.7	48.65 ± 7.55*
KC	23.8 ± 2	nd
TGF-β	19063.55 ± 4377.45	nd
BALf		
IL-2	36.6 ± 0.7	38.4 ± 0.2
IFN-γ	144 ± 27.5	nd
TNF-α	20.4 ± 0.8	5.2 ± 0.6*
IL-10	8.7 ± 0.6	2.15 ± 0.15*
IL-4	94.75 ± 9.35	nd
IL-5	94.85 ± 13.35	10.1 ± 0.8*
IL-13	131.8 ± 3.9	9.1 ± 8.3*
Eotaxin	431 ± 84.7	12 ± 0.2*
MMP-9	945.3 ± 23.9	371.05 ± 67.25*
SDF-1α	260.8 ± 17.5	35.75 ± 2.95*
KC	17.9 ± 0.6	6.95 ± 0.45*
TGF-β	467 ± 15.7	nd

The concentration of cytokines was measured by outsourcing to Pierce, Endogen by multiplexing with their patented technology. Presented data evaluation was outsourced to Pierce Endogen. Searchlight™ technology was used to assay the following cytokines in triplicate and data presented is mean of samples from five mice per group and pooled from two independent experiments ± SEM. Other IL-2 all Th1 and Th2 as well as other associated cytokines levels were decreased in KO BALf while a universal downregulation was found in all cytokines in plasma. Presented data is average of triplicate of samples ($n = 5$/group) ± SEM of two independent experiments

nd not detected

*p value < 0.01 compared to WT + alum (baseline) values

Adoptive Transfer of CD62L + CD4+ Splenocytes from Sensitized WT to Naïve ELP−/− Recipients Could Reverse Inhibition of Asthma Development

Our next question was if similar to what we found in all the previous quantifications, in the deleted scenario too allergic asthma would fail to develop or to reverse the effects of deletion, and if so, then which cells would be most effective so that we may clearly delineate the role playing of that particular cell subset that is responsible for the pathology by virtue of their expression of the triple selectins. To this end, we did the following elimination experiment by various permutations and combinations. Our most obvious candidate was the CD4+ T cells. To assess whether adoptively transferred CD4+ T cells from sensitized wild-type mice, where all three selectins are in place, could reverse the nondevelopment of asthma phenotype in ELP−/− mice, adoptive transfer of the aforementioned cells from spleen of sensitized WT mice was transferred by tail vein injection to

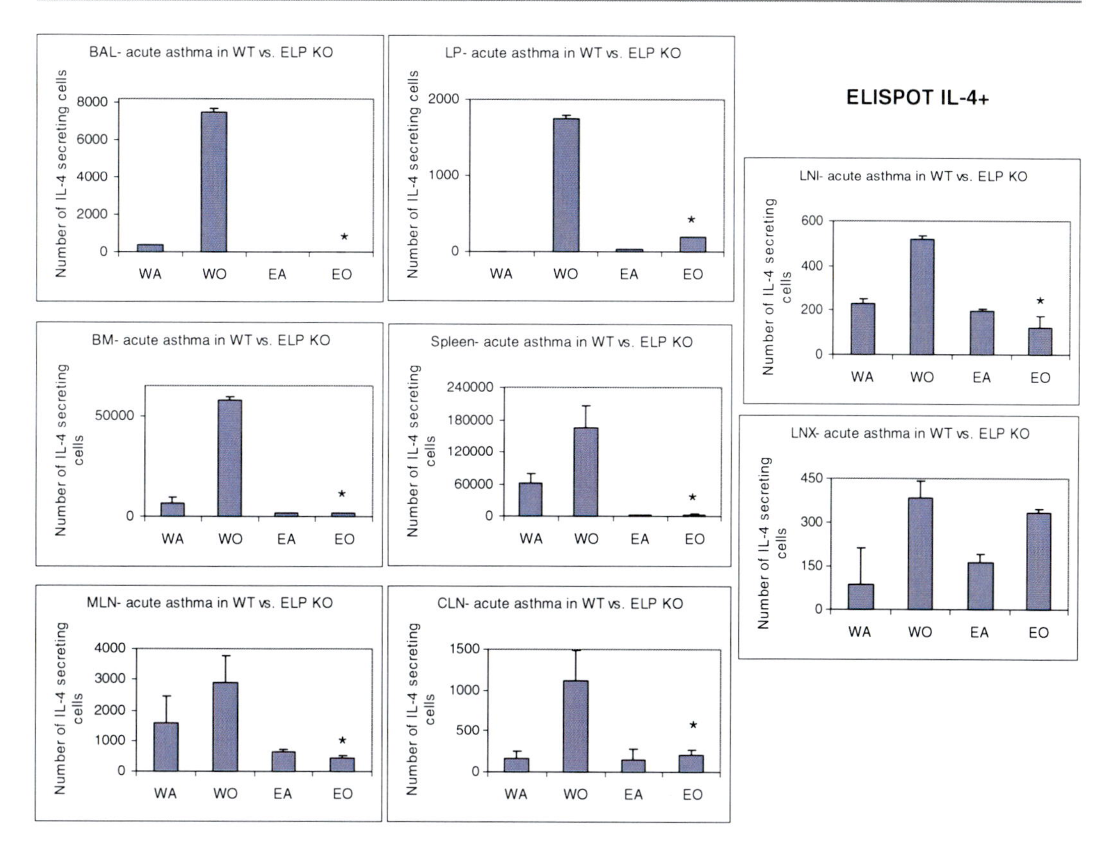

Fig. 3.20 ELISPOT of Th2 cells in lymphoid tissues cells. Number of IL-4+ cells averaged from triplicate wells of a 96-well ELISPOT plate ± SEM of 2 independent experiments.* denotes p value < 0.01 compared to WT + OVA

naïve ELP−/− mice followed by OVA challenge as described in "Materials and Methods." Surprisingly, we found that these ELP−/− mice did respond to the OVA challenge and developed a significant asthma phenotype (Gr#2) compared to either the Grs# 3 or 4 where CD4+ T cells from spleen of sensitized ELP−/− mice were used. The experiment was validated by Gr#1 (positive control) (Table 3.6 and Fig. 3.23). Functionally incompetent dendritic cells in knockout mice ruled out. We also wanted to check whether the regulatory T cells and dendritic cells were similar in the knockout mice. Table 3.7 shows that both WT and KO tissues showed similar expression of Treg and dendritic cells.

Discussion

A similar protective effect, as was observed in α4−/− and β2−/− mice in our earlier studies in the acute allergic asthma model, was also observed here with the ELP−/− mice [109]. This raises the question of redundancy of the roles of all these cell adhesion molecules, each cardinal in their individual capacities in cellular migration in inflammation and each affecting specific cellular components in the inflammatory cascade. So the fulcrum of the following discussion in an effort to interpret the observations described in this paper shall revolve around (a) the role played by all three selectins in context with the cells involved in the onset, development, and maintenance of allergy and associated inflammation in acute asthma, (b) their role in the etiology of the disease or cooperative functions at later stages of the disease manifestation, and (c) whether pathways traversed by these adhesion molecules intersect, overlap at certain points, or are indeed redundant. We shall try and address each point in our discussion of the data in the following paragraphs.

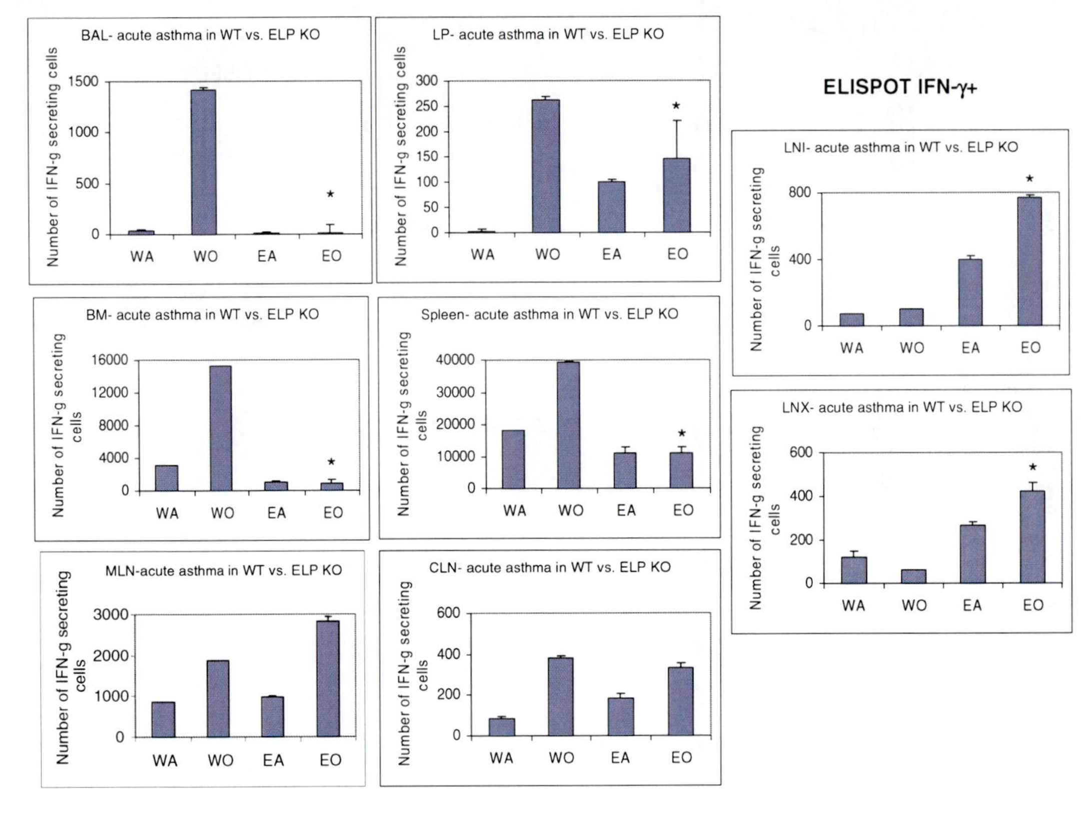

Fig. 3.21 ELISPOT of Th1 cells in lymphoid tissues cells. Numbers of IFγN + -cells averaged from triplicate wells of a 96-well ELISPOT plate ± SEM of 2 independent experiments. *denotes p value < 0.01 compared to WT + OVA

First of all, as the data in our study shows, despite increased and extensive peripheral leukocytosis, especially in eosinophils, deletion of all three selectins leads to a complete inhibition of the development of the asthma phenotype in the OVA-treated ELP−/− mouse. The failure of the eosinophils to travel firstly to the lung and thence to the airways to elicit the downstream effects of composite asthma phenotype may have been abrogated due to the absence of either E- or L-selectin or all three. Let us try and eliminate the possible candidates.

Previous studies [110] have shown a key role for P-selectin in modulating leukocyte behavior, e.g., ragweed-induced recruitment of eosinophils to the peritoneum suggests a crucial role for P-selectin in Eos recruitment. Neither E- nor L-selectin appears to mediate leukocyte recruitment in TNF-α-induced inflammation or thioglycollate-induced peritonitis. We should keep in mind, however, that all three selectins cooperate at some level to influence eosinophil homeostasis. While P-selectin is the only selectin whose absence impairs the recruitment of these cells to the inflamed peritoneum, the combined absence of P- and E-selectins seems to lead to the complete abrogation of the allergic response into the development of an asthma phenotype [111].

However, the conclusive role of L-selectin versus all three is somewhat controversial. Previous works by different groups have presented somewhat disparate results. A study with an anti-L-selectin antibody or L-selectin gene knockout mice could not prevent development of asthma [112]. Other groups have shown that neither L- nor E-selectin contributed to sinusoidal neutrophil sequestration or transmigration. Yet other groups have shown that adoptive transfer of splenic lymphocytes from cockroach antigen (CRA)-primed E- and

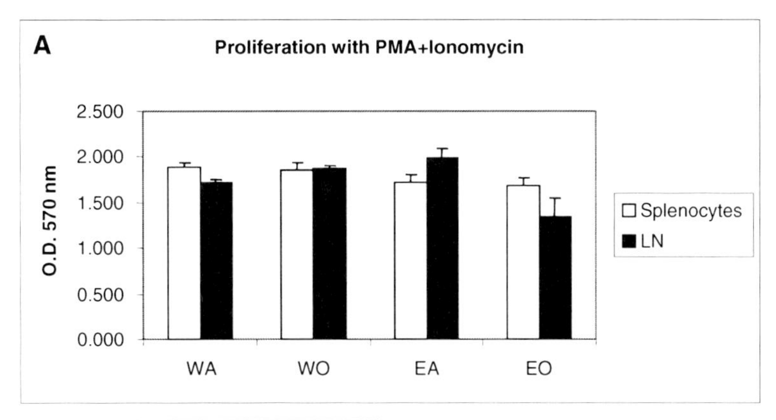

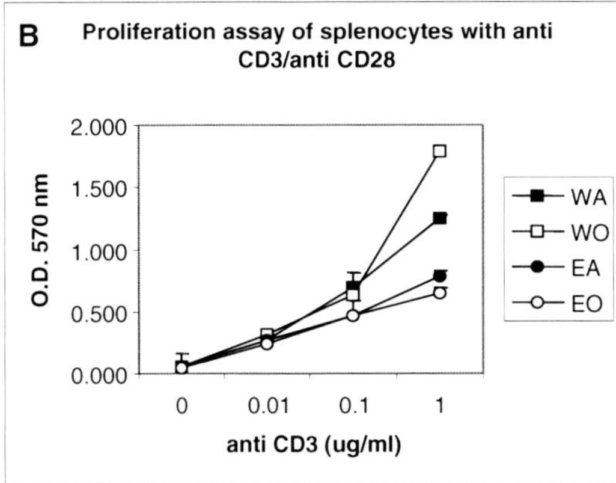

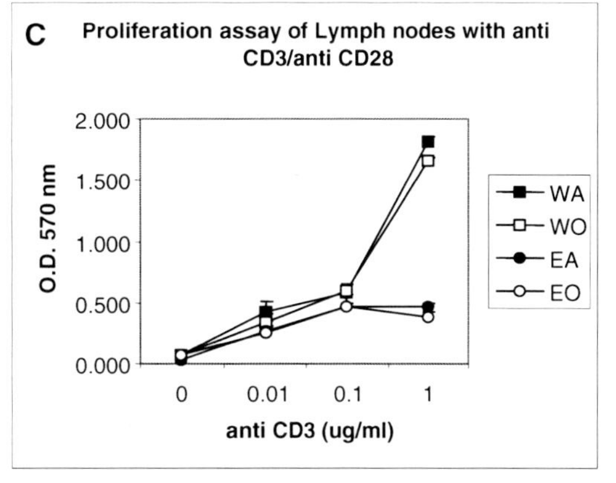

Fig. 3.22 Functional tests for T cells. Unseparated T cells from spleen and peripheral lymph nodes (pooled cervical and axillary lymph nodes) were evaluated for proliferative responses to various stimuli and readout taken at 570 nm following MTT assay (Promega) (**a**) Proliferation assay with PMA + ionomycin; (**b**) Proliferation to anti CD3/CD28 with cells from spleen, and (**c**) with cells from the same lymph nodes as above

P-selectin-deficient cells into naïve wild-type (WT) mice produced the same level of airway hyperreactivity as transfers from CRA-primed WT into naïve WT hosts, indicating similar peripheral immunization and [113] similar serum IgE production of the selectin-deficient and WT animals, indicating that the Th2-driven isotype switch was unaffected by the genetic alterations. Thus both P- and E-selectin contribute to CRA-induced peribronchial inflammation and airway hyperreactivity [100]. Exclusive role of L-selectin in OVA asthma was shown by the following studies: While OVA-sensitized/challenged ICAM-1-deficient mice showed decreased levels of β220+ lymphocytes in the BALf, OVA-sensitized/challenged L-selectin-deficient mice demonstrated significantly reduced numbers of CD3+ lymphocytes and increased β220+ lymphocytes in BALf, suggesting a crucial role for ICAM-1 in airway inflammation and AHR in asthma (as corroborated by our earlier studies [92, 93]) but L-selectin plays a more selective role in the development of AHR but independent of airway inflammation in this animal model of asthma [114]. Thus earlier studies have shown that both cell adhesion molecules (i.e., selectins and integrins) play key roles in cell trafficking and in the lung they regulate leukocyte extravasation, migration within the interstitium, cellular activation, and tissue retention.

However, adoptive transfer experiments show very clearly that sensitized CD4+ T cells from WT spleens were able to reverse the effect of ELP deletion in developing an asthma phenotype

Table 3.6 Adoptive transfer groups

	Donors	Recipients	Asthma phenotype
Gr#1	CD4+ from sensitized WT	Nave WT	+
Gr#2	CD4+ from sensitized WT	Nave ELP–/–	+
Gr#3	CD4+ from sensitized ELP–/–	Nave WT	–
Gr#4	CD4+ from sensitized ELP–/–	Nave ELP–/–	–

Donors were always sensitized mice and CD4+ T splenocytes were isolated by MACS and adoptively transferred into nave recipients (5×10^6) by tail vein injection. Splenocytes were collected a week after sensitization of donor by i. p. injection of OVA-alum. The recipients were then challenged by intratracheal OVA instillation over the next 72 h and sacrificed and evaluated 24 h after the last challenge

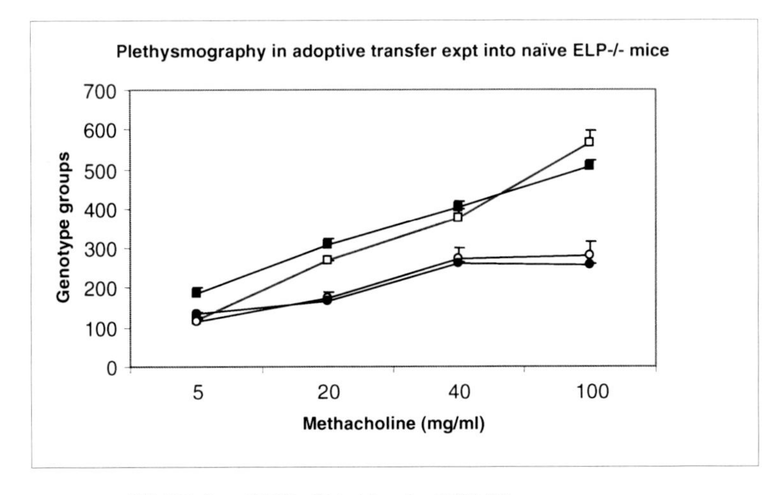

Fig. 3.23 Noninvasive whole body plethysmography to assess pulmonary function in adoptively transferred recipients. ($n = 4$ per group). [Refer Table 3.3 for adoptive transfer groups.] Gr#1 (*white square*), Gr#2 (*black square*), Gr#3 (*white circle*), Gr#4 (*black circle*)

although of a smaller magnitude than sensitized and challenged WT. It follows that CD4+ T cells from WT will express only L-selectin of all the three selectins, and therefore, L-selectin alone was sufficient to reverse the effects of E-, L-, and P- selectin deletion [115]. The significant reduction in levels of all immunoglobulins (Fig. 3.19) indicates a serious problem with Ig sequestering by ELP–/– B cells. If CD62L + CD4+ T cells from sensitized WT spleen could develop asthma in naïve ELP–/–, it also indicates that development of asthma may not be entirely dependent on ELP–/– B cells in the spleens of these ELP–/– recipients or that the five million CD62L + sensitized CD4 + T cells are sufficiently potent to go into circulation and generate at least threshold amount of IgE for required airway inflammatory response.

As seen from ELISPOT assays (Fig. 3.20), only the axillary lymph nodes showed increase in IL-4 + –expressing cells which may have provided the required mobilizing force for the nearby airway inflammation in the adoptively transferred recipients. The role of selectins in neutrophil trafficking in the lungs was frequently considered negligible since the narrow pulmonary capillaries cannot accommodate the typical selectin-mediated rolling phenomenon.

Furthermore, selectins did not seem necessary in lung neutrophil sequestration since the deceleration of circulating neutrophils prior to their firm adherence was effectively achieved by their mechanical retention [91]. Yet a large body of experimental data demonstrating that selectin inhibition (via the use of blocking antibodies or selectin antagonists or transgenic knockout

Table 3.7 Number of total and LECAM + regulatory T cells and dendritic cells in various hematopoietic and non-hematopoietic tissues in WT versus KO mice

	Total CD45+	CD3+	CD4+	CD8+	CD4+ CD25+	CD11c+
BM						
WO	9.67 ± 2.31	0.15 ± 0.06	0.02 ± 0.001	0.12 ± 0.09	0.19 ± 0.06	0.020 ± 0.004
EO	10.16 ± 3.34	0.18 ± 0.07	0.03 ± 0.015	0.14 ± 0.03	0.18 ± 0.07	0.018 ± 0.006
PB						
WO	26.90 ± 4.41	9.49 ± 4.76	3.59 ± 1.05	5.90 ± 0.7	1.38 ± 0.56	1.18 ± 0.76
EO	29.99 ± 5.67	6.49 ± 2.13	2.22 ± 1.02	4.27 ± 0.56	1.05 ± 0.74	1.61 ± 0.13
Spleen						
WO	234.89 ± 54.87	118.69 ± 32.87	48.79 ± 11.09	69.90 ± 11.21	27.01 ±	53.91 ± 4.98
EO	212.73 ± 43.98	109.09 ± 18.24	44.76 ± 4.97	63.33 ± 14.32	30.73 ±	41.61 ± 9.61
MLN						
WO	38.74 ± 4.73	15.22 ± 5.52	4.85 ± 14.09	10.37 ± 2.21	4.95 ± 0.54	7.00 ± 0.43
EO	36.10 ± 9/65	17.48 ± 5.98	4.95 ± 0.67	12.53 ± 3.86	5.30 ± 0.04	7.42 ± 1.09
CLN						
WO	16.35 ± 3.65	4.41 ± 1.06	1.452 ± 0.04	2.89 ± 0.03	12.91 ± 0.54	2.78 ± 0.54
EO	14.00 ± 4.45	6.36 ± 0.64	5.05 ± 0.43	1.31 ± 0.54	3.51 ± 1.85	2.68 ± 0.31
LNI						
WO	168.37 ± 7.54	8.67 ± 2.32	2.76 ± 0.54	5.91 ± 1.09	2.82 ± 0.06	3.98 ± 0.43
EO	21.75 ± 3.85	10.53 ± 0.65	2.980.54 ±	7.55 ± 2.03	3.19 ± 1.07	4.47 ± 1.14
LNX						
WO	11.90 ± 3.28	5.61 ± 0.43	3.93 ± 0.54	3.68 ± 0.54	2.43 ± 0.87	2.26 ± 1.09
EO	13.06 ± 6.90	5.94 ± 1.33	4.71 ± 2.44	2.82 ± 0.98	3.28 ± 0.54	2.50 ± 0.23
LP						
WO	0.70069 ± 0.01	0.01962 ±0.0030.004	0.00294 ± 0.003	0.01668 ± 0.003	0.09459 ± 0.003	0.00090 ± 0.00001
EO	0.71086 ± 0.04	0.01987 ±	0.00327 ± 0.001	0.01660 ± 0.004	0.09761 ± 0.004	0.00092 ± 0.00004

Cell suspension was prepared either by flushing or homogenizing tissues such as bone marrow (BM) cells were prepared by flushing femurs, PB (peripheral blood cells were prepared by hemolyzing whole blood to remove RBCs, LP (lung parenchyma) cells were prepared by homogenizing exsanguinated lung tissue, BALf (bronchoalveolar lavage fluid) cells were prepared by flushing airways in the exsanguinated lung, spleen cell suspension was prepared by flushing the spleen and MLN (mesenteric lymph node), LNI (inguinal lymph node), CLN (cervical lymph node) and LNX (axillary lymph node) cells were prepared by similar method. The cells were resuspended in 1X PBS with 5 % BSA and stained with fluorochrome-conjugated antibodies and quantified by flow cytometry. WO denotes wild-type mice challenged with OVA, EO denotes ELP−/− challenged with OVA, and EO denotes ELP−/− challenged with OVA. Presented data is average of triplicate of samples ($n = 5$/group) ± SEM of 2 independent experiments

*p value < 0.01 compared to WT + alum (baseline) values

of one or more selectins) frequently protected animals from acute lung Injury [95]. In the light of the above data, let us try and explain what the status of the various cells playing key role in the onset and establishment of the TH2 inflammatory response to the allergen OVA is likely to be. First, the dendritic cells [116] and alveolar macrophages [117], whose responses are critical for antigen presentation to the naïve T cells, need P- and L-selectins, and the β2 integrins and L-selectins respectively may be considered. These cells therefore are practically dysfunctional when L-selectins are absent on the KO mice. They are neither capable of processing the signal nor coupling with their other accessories that mediate adhesion in the inflammatory cascade. If we assume that in the ELP−/− mice, there was no response to the allergen in the first place because the cells themselves were incapable of sequestering the allergen for presentation, the absence of any TH2 response and therefore inflammatory recruitment downstream and the consequent non-manifestation of the composite asthma phenotype may be explained. However, (Table 3.4) increased number of mature leukocytes and progenitors in BM and circulation of post-OVA ELP−/− mice but not so in their WT counterparts show that for stromal attachment of progenitors in the BM and release into circulation, presence of selectins is a critical signal. This is in significantly so in the airways But elevated in the lung parenchyma particularly in the number of monocytes, macrophages, and neutrophils (myeloid population) but no eosinophils and mast cells. This is somewhat confusing because it seems that the systemic response did happen to the OVA challenge but migration to the tissues was arrested.

The adoptive transfer of only ELP + CD4+ T lymphocytes reinstating the asthma phenotype indicates conclusively that it was the afferent arm of the Th2 immune response that was affected. However, since human asthma involves a complex interplay of cells, early and late asthma responses sometimes intermingle especially during exacerbations in the chronic phase of allergic asthma, hence the assumption that the afferent limb of the immune response was likely affected. Also, since the role of CD4+ T cells in eosinophil recruitment is firmly established [118–120], triple selectin deletion may be held responsible for preferably affecting the afferent arm of the response by giving the OVA-treated mouse such a clean respiratory tree and no hyperreactivity to methacholine.

As for attenuation of both Th1 and Th2 cytokines in the ELP−/− OVA-treated mice, the nonoccurrence of the first lap of the journey (the afferent limb of the immune response) probably prevents activation and recruitment of the cytokine secreting cells [121]. The elegant study by David A. Randolph et al [122] shows that while both Th1 and Th2 cells are recruited to the airways, Th1 cells predominate early and Th2 cells predominate late. On this assumption we conclude that it is not exclusively the Th2 cells that are recruited and further reduction in Th1 and Th2 cytokine secretion correlates with the complete nonoccurrence of the first line response in the development of the composite asthma phenotype in the knockout animals. In the light of our findings [92, 93] with the acute and chronic OVA models, the β2 integrins, which, compared to α4 integrins, preferably co-localize with the P- and L-selectins in these cells, there seems to be a disruption of a similar pathway that also prevents development of asthma in the CD18 knockout mice.

However, release of Th2-specific cytokines being significantly downregulated in KO BALf and plasma (Table 3.5) and virtually no IL-4-secreting cells being detected in all tissues systemic as well as local (Tables 3.5 and 3.6), a universal lack of response in processing the stimulus, viz., OVA, is indicated. The cells that were hyperproliferated in bone marrow and blood were also incapable of sequestering the requisite cytokines to mount a response. Interestingly peripheral lymph nodes release higher levels of TH1 cytokines indicating that deletion of all three selectins may be critical for skewing the inflammatory responses toward a TH2 phenotype at least in an acute allergic setup. Whether the situation will reverse in the chronic phase as in our work with the β2 integrin knockout mice [93] is yet to be studied.

This is the first report that clearly delineates the role of L-selectin from the other selectins and shows that (a) ELP−/− mice are incapable of mounting a full-fledged asthma response despite impressive peripheral leukocytosis possibly due to a failure to sequester initial response to allergen exposure; (b) inflammatory response in airways is almost nonexistent despite some myeloid migration into their lungs; (c) ELP−/− T cells are functionally active and responsive, whereas ELP−/− B cells are incapable of sequestering total as well as allergen-specific antibody which may indicate (1) that the B cells to be "primed" during initial sensitization by OVA-alum complex may have been ineffective (which is more likely) or (2) the primed B cells when matured into plasma cells fail to properly sequester OVA-specific IgE (which, although an attractive explanation, cannot explain how in the adoptive transfer experiments (Fig. 3.23 and Table 3.6), naïve ELP−/− recipients could successfully mount an asthma response when transplanted with CD4+ WT T cells which could happen only if the sensitization phase occurred successfully). Nevertheless the fact that they were able to manifest "functional asthma" (as shown by lung function test) despite unprimed B cells may indicate that the robust response was due mainly to the CD4+ T-cell response that was equipped with the L-selectin and the endothelium of the challenged mouse having their E-selectins intact and, lastly, P-selectins on the recipient platelets which may have played a key role in neutrophil sequestration (Tables 3.2, 3.3, and 3.4); (d) sensitized CD62L + CD4+ T cells adoptively transferred into naïve ELP–/– recipients could reverse the effects of ELP deletion and nondevelopment of asthma; and lastly (e) the axillary lymph nodes may be key to mobilize IL-4+ cells/IL-4 for Th2 response and inflammatory recruitment in the airways, the significance of which at this point is unclear.

Due to the essential role played by these cell adhesion molecules in lung inflammation, all selectin family members (including L-selectin, P-selectin, and E-selectin) appear to be important therapeutic targets [123]. Indeed, Revotar Biopharmaceuticals AG is developing the drug under license from Encysive, which is a CD62L antagonist (TBC-1269) for the potential treatment of asthma, COPD, VILI, ALI, and ARDS.

Conclusion

Comparative studies between ELP–/– and WT mice uncovered functional roles of these three integrins in inflammatory response in allergic asthma. All three selectins seem to impede inflammatory migration, while only L-selectin also possibly regulates activation of specific T-cell subsets in the lung and airways. Also, E-, L-, and P- selectins are important drug targets for asthma. Locally applicable combination therapies with small molecular antagonists or antibodies may be useful in acute allergic asthma treatment regimens to supplement other conventional therapies.

Materials and Methods

Animals

C57BL6 mice were used as described previously [92, 93]. Wild type (WT) and ELP–/– (was kindly donated by Richard O. Hynes of the Howard Hughes Medical Institute and Center for Cancer Research, MIT, Cambridge, MA) both on a C57BL6 background were used. All animals were maintained under SPF conditions in the animal facility of the University of Washington following strict guidelines laid down by IACUC ($n = 5$ mice per experimental group). Allergen sensitization and challenge mice were sensitized and later challenged with OVA (Pierce, Rockford, IL) as described previously [92, 93]. Mice were immunized with OVA (100 mg) complexed with aluminum sulfate in a 0.2-ml volume, administered by i.p. injection on day 0. On days 8 (250 mg of OVA) and on days 15, 18, and 21 (125 mg of OVA), mice were anesthetized briefly with inhalation of isoflurane in a standard anesthesia chamber and given OVA by intratracheal (i.t.) administration. Intratracheal challenges were done as described previously (Iwata A, J Immunol. 2003;170:3386). Mice were anesthetized and placed in a supine position on the board. The animal's tongue was extended with lined forceps, and 50 ml of OVA (in the required concentration) was placed at the back of its tongue. The control group received normal saline with aluminum sulfate by i.p. route on day 0 and 0.05 ml of 0.9 % saline by i.t. route on days 8, 15, 18, and 21 (Fig. 3.17).

Pulmonary Function Test

In vivo airway hyperresponsiveness to methacholine was measured 24 h after the last OVA challenge by both invasive and noninvasive plethysmography.

Invasive Plethysmography

On day 22, 24 h after the last intratracheal allergen (OVA) challenge, invasive pulmonary mechanics were measured in mice in response to methacholine in the same manner as previously described [92] with the following modifications: (a) the thorax was not opened, (b) mice were ventilated with a tidal volume of 200 μl and respiratory rate of 120 breaths/min using a MiniVent Ventilator for Mice (Harvard Apparatus, Holliston, MA), (c) mice received aerosolized solutions of methacholine (0, 3.125, 6.25, 12.5, 25, 50, and 100 mg/ml in normal saline) via an

AER 1021 nebulizer aerosol system (Buxco Electronics, Inc., Wilmington, NC) with 2.5–4 μm aerosol particle size generated by NEB0126 nebulizer head (Nektar Therapeutics, San Carlos, CA), and (d) a commercial plethysmography system (Model PLY4111 plethysmograph, MAX II amplifier and pressure transducer system, and Biosystem XA software, Buxco Electronics, Inc.) was used to determine RL as calculated from measures of pressure and flow and expressed as $cmH_2O/ml/s$. Noninvasive plethysmography (expressed as Penh) was also assessed on day 22 in independent experiments.

Noninvasive Whole-Body Plethysmography

Noninvasive whole-body plethysmography was used to measure lung function in conscious, free-moving, spontaneously breathing mice using whole-body plethysmography (model PLY 3211; Buxco Electronics, Sharon, CT) as previously described [116]. Mice were challenged with aerosolized saline or increasing doses of methacholine (5, 20, and 40 mg/ml) generated by an ultrasonic nebulizer (DeVilbiss Health Care, Somerset, PA) for 2 min. The degree of bronchoconstriction was expressed as enhanced pause (Penh), a calculated dimensionless value, which correlates with the measurement of airway resistance, impedance, and intrapleural pressure in the same mouse. Penh readings were taken and averaged for 4 min after each nebulization challenge. Penh was calculated as follows: Penh = [(Te/Tr-1)X (PEF/PIF)], where Te is expiration time, Tr is relaxation time, PEF is peak expiratory flow, and PIF is peak inspiratory flow × 0.67 coefficient. The time for the box pressure to change from a maximum to a user-defined percentage of the maximum represents the relaxation time. The Tr measurement begins at the maximum box pressure and ends at 40 %.

Cell Suspensions Prepared from Tissues

After pulmonary function testing, the mouse underwent exsanguination by intraorbital arterial bleeding and then BALf (0.4 ml three times) of both lungs. Total BALf fluid cells were counted from a 50-ml aliquot, and the remaining fluid was centrifuged at 200 g for 10 min at 4 °C, and the supernatants were stored at −70 °C for assay of BALf cytokines later. The cell pellets were resuspended in FCS and smears were made on glass slides. The cells, after air-drying, were stained with Wright-Giemsa (Biochemical Sciences Inc, Swedesboro, NJ), and their differential count was taken under a light microscope at 40X magnification. Cell number refers to that obtained from lavage of both lungs/mouse. Lung parenchyma was prepared in the following way: lung mincing and digestion was performed after lavage as described previously [46] with 100 u/ml collagenase for 1 h at 37 °C and filtered through a 60# sieve (Sigma). All numbers mentioned in this paper refer to cells obtained from one lung/mouse.

Lung Histology

Lungs of other animals of same group were fixed in 4 % paraformaldehyde overnight at 4 °C. The tissues were embedded in paraffin and cut into 5-mm sections. A minimum of 15 fields were examined by light microscopy. The intensity of cellular infiltration around pulmonary blood vessels was assessed by hematoxylin and eosin staining. Airway mucus was identified by staining with Alcian blue and periodic acid Schiff staining as described previously [92, 93].

Smear Evaluation

Proportions of eosinophils and mast cells were assessed in cytospin smears stained with hematoxylin and eosin by Diff Quik stain from Fisher. Fluorescein-activated cell sorter (FACS) analysis cells from hemolyzed peripheral blood (PB), bone marrow (BM), bronchoalveolar lavage (BALf), lung parenchyma (LP), and spleen were analyzed on a FACSCalibur (BD Immunocytometry Systems, San Jose, CA) by using the CellQuest program. Staining was performed by using antibodies conjugated to fluorescein isothiocyanate (FITC), phycoerythrin (PE), allophycocyanin (APC), peridinin-chlorophyll-protein (Per CP-Cy5.5), and Cy-chrome (PECy5 and PE-Cy7). The following BD Pharmingen (San Diego, CA) antibodies were used for cell-surface staining: APC-conjugated CD45 (30 F-11), FITC-conjugated CD3 (145-2C11), PE-Cy5-conjugated CD4 (RM4-5), PE-conjugated CD45RC (DNL-1.9), APC-conjugated CD8(53–6.7), PE-Cy5-conjugated β220 (RA3-6β2), FITC-conjugated IgM, PE-conjugated CD19 (ID3), PE-conjugated CD21 (7G6), FITC-conjugated CD23 (B3B4), APC-conjugated GR-1(RB6-8C5), and PE-conjugated Mac1(M1/70). PE-Cy5-conjugated F4/80 (Cl:A3-1(F4/80)) was obtained from Serotec Ltd., Oxford, UK.

CFU-C Assay

To quantitate committed progenitors of all lineages, CFU-C assays were performed using methylcellulose semisolid media (Stemgenix, Amherst, N.Y.) supplemented with an additional 50 ng of stem cell factor (Peprotech, Rocky Hill, N.J.) per ml. Next, 50,000 cells from bone marrow, 500,000 cells from spleen, 0.01 million cells from lung and BALf, and 10 ml peripheral blood were plated on duplicate 35-mm culture dishes and then incubated at 37 °C in a 5 % CO_2-95 % air mixture in a humidified chamber for 7 days. Colonies generated by that time were counted by using a dissecting microscope, and all colony types (i.e., burst-forming units-erythroid [BFU-e], CFU-granulocyte-macrophage [CFU-GM], and CFU-mixed [CFU-GEMM]) were pooled and reported as total CFU-C. Total CFU-C per organ was calculated by extrapolating CFU-C against number of plated cells to the total number of cells in the organ.

ELISA for Cytokines

Th2 cytokines (IL-4 and IL-5) and TNFa and IFNγ in BALf and serum (previously frozen at −70 °C) were assayed with mouse Thl/Th2 cytokine CBA (BD Biosciences, San Diego, CA) following the manufacturer's protocol. According to the manufacturer's protocol, IL-13 and eotaxin were measured by Quantikine M kits from R&D Systems, Minneapolis, MN.

ELISPOT

IL-4+ and IFN-γ+ cells in single-cell suspensions from lung parenchyma and BALf were detected employing standard ELISPOT assays (Lee S-H, Nat Med. 2003;9:1281) using detection and capture monoclonal antibodies and AEC substrate reagent from BD Biosciences. Dots were counted manually using 40X magnification.

OVA-Specific IgE and IgG1 in Serum

Anti-mouse IgE (R35-72) and IgG1(A85-1) from BD Biosciences, San Diego, CA, were used for measuring OVA-specific IgE and IgG1 (in serum previously frozen at −70 °C) respectively by standard ELISA procedures as previously described [57].

T-Cell Proliferation Assay

MACS-separated CD4+ and CD8+ T cells from spleens were stimulated in vitro with various concentrations of stimuli (CD3/CD28, phorbol myristic acetate (PMA)/ionomycin, irradiated antigen-presenting cells (APCs), and lipopolysaccharide (LPS)) to assay proliferative responses. After 72 h, proliferation was measured by CellTiter96 assay from Promega (Madison, WI, USA) measuring OD at 570 nm.

Adoptive Transfer

At day 8 after i.p. sensitization, 5 × 106 CD4+ splenocytes from both WT controls or ELP−/− mice were purified by magnetic-activated cell sorting (MACS, Miltenyi Biotec, Auburn, CA, USA) and then injected into the tail veins of naïve controls or ELP−/− recipients. The mice were subsequently challenged with 3 i.t. instillations of OVA over the next 72 h and sacrificed 24 h after the last instillation [58]. For negative control (Gr#4), CD62L-CD4+ T cells were separated by MACS from ELP−/− donors. Gr#1 is the positive control here.

Statistical Analysis

Statistical differences among samples were tested by Student's t-test. P value less than 0.05 was considered statistically significant.

Subchapter 3: Role of gp91phox Subunit of NADPH Oxidase and MMP-12 in an Acute Inflammatory and an Acute Degenerative Pulmonary Disease Model Using Genetic Knockout Mice

A. **Role of gp91phox subunit of NADPH oxidase and MMP-12 in acute allergic mouse asthma model**
B. **Role of gp91phox subunit of NADPH oxidase and MMP-12 in an acute degenerative mouse model of bleomycin-induced pulmonary fibrosis**
C. **Specific role of TH2 cells in gp91phox regulated acute allergic response**

Summary of the Study

Objective: For establishment of inflammation, a constant interplay between different effector cells from circulation, local resident cells, soluble mediators, and genetic host factors is required. Molecular mechanisms, initiating and perpetuating inflammation, in particular, the involvement of effector cells in redox reactions for producing O_2^- (superoxide anion) through the mediation of NADPH oxidase is a critical step. Prior data suggest that reactive oxygen species (ROS) produced by NADPH oxidase homologues in non-phagocytic cells play an important role in the regulation of signal transduction, while macrophages use a membrane-associated NADPH oxidase to generate an array of oxidizing intermediates which inactivate MMPs on or near them.

Materials and Methods and Treatment: To clarify the role of NADPH oxidase in T-cell-initiated, macrophage-associated allergic asthma, we induced allergen-dependent inflammation in a gp91^{phox}–/– mouse.

Results and Conclusion: Both inflammation and airway hyperreactivity were more extensive than in wild-type mice post-OVA. Although OVA-specific IgE in plasma were comparable in wild-type and knockout mice, enhanced inflammatory cell recruitment from circulation and cytokine release in the lung and BALf, accompanied by higher airway resistance as well as Penh in response to methacholine, indicate a regulatory role for NADPH oxidase in development of allergic asthma. While T-cell-mediated functions like Th2 cytokine secretion and proliferation to OVA were upregulated synchronous with the overall robustness of the asthma phenotype, macrophage upregulation in functions such as proliferation, mixed lymphocyte reaction, and MCP-1 directed chemotaxis, but downregulation of respiratory burst response, indicates a forking in their signaling pathways. gp91^{phox}–/– MMP12 double knockout (DKO) mice show a similar phenotype as the gp91^{phox}–/– showing the noninvolvement or synergistic involvement of MMP12 in the response pathway. In mixed lymphocyte reaction using the increased B7.1 but reduced B7.2 and MHC class II expression indicating alteration of costimulatory molecule expression critical for T-cell activation on both gp91^{phox}–/– and DKO mice may explain the mechanism by which gp91^{phox} may regulate Th2 pathway in allergic asthma.

Introduction

Asthma is a complex syndrome with a well-described pathology. However, animal and clinical studies in humans continue to provide conflicting data on contribution of local cells, viz., airway epithelial, endothelial and smooth muscle cells, and fibroblasts, versus cells recruited from circulation. Asthma is characterized by accumulation of inflammatory cells in the lung and airways, secretion of predominantly Th2 cytokines in the lung and airways, epithelial desquamation, goblet cell hyperplasia, mucus hypersecretion, and thickening of submucosa resulting in bronchoconstriction and airway hyperresponsiveness. Dysregulated immunity seems to suppress Th1 response and triggers Th2 response whose development is promoted by antigen-presenting cells. Th2 cytokines (IL-4, IL-5, IL-9, IL-13) were quantitated from these cells of which IL-4 and IL-13 promote B-cell differentiation into plasma cells that secrete IgE. Cross-linking of IgE receptors on mast cells releases histamines, prostaglandins, thromboxane, and leukotrienes, leading to bronchoconstriction, vasodilation, and mucus secretion. A cascade of interactions between cells and soluble molecules result in bronchial mucosal inflammation and lead to airway hyperresponsiveness.

The production of superoxide anions (O_2^-) by neutrophils and other phagocytes is an important step in our body's innate immune response. O_2^- is the precursor of a range of chemicals generally referred to as ROS (reactive oxygen species) [124]. These act as microbicidal agents and kill invading microorganisms either directly or through the activation of proteases [125, 126]. O_2^- is produced by the NADPH oxidase, a multi-protein enzyme complex, which is inactive in resting phagocytes, but becomes activated after interaction of the phagocyte with pathogens and their subsequent engulfment in the phagosome [127]. Defects in the function of the NADPH oxidase result in a severe immunodeficiency, and individuals suffering from CGD (chronic granulomatous disease), a rare genetic disorder that is caused by mutations in NADPH oxidase genes, are highly susceptible to frequent and often life-threatening infections by bacteria and fungi [128]. The microbicidal activity of ROS has generally been seen as the only beneficial function of these chemicals, and uncontrolled production of ROS has been implicated in tissue destruction and a number of disease states. However, over the last couple of years, it has become apparent that ROS produced by NADPH oxidase homologues in non-phagocytic cells also play an important role in the regulation of signal transduction, often via modulation of kinase and phosphatase activities or through gene transcription [129]. These NADPH oxidase homologues are referred to as Nox enzymes (gp91^{phox} is specified as Nox2, where phox is *ph*agocytic *ox*idase), and several members of this novel protein family have been identified so far.

There is increasing evidence that redox regulation of transcription, particularly activator protein-1 (AP-1) and nuclear factor kappa B (NF-κB), is important in inflammatory diseases. NADPH oxidase, the primary source of reactive oxygen species, is a strong candidate for the development of therapeutic agents to ameliorate inflammation and end-organ damage. The possibility of gene therapy for inherited diseases with a single gene mutation had been verified by the successful treatment with bone marrow transplantation. As the gene therapy method and theory has been progressing rapidly, it is expected that gene therapy will overcome the complications of bone marrow transplantation. Of these inherited diseases, chronic granulomatous disease (CGD) is one of the most expected diseases for gene therapy. CGD is an inherited immune deficiency caused by mutations in any of the following four phox genes encoding subunits of the superoxide-generating phagocyte NADPH oxidase. It consists of membranous cytochrome b558 composed of gp91 phox and p22 phox and four cytosolic components, p47 phox, p67 phox, rac p21, and p40 phox, which translocate to the membrane upon activation. The gp91^{phox} subunit (also called the β-subunit of the cytochrome) consists of 570 amino acids and has a molecular mass of 65.3 kDa, but runs as a broad smear of approx. 91 kDa on SDS/polyacrylamide gels due to a heterogeneous glycosylation pattern of three asparagine residues (Asn^{132}, Asn^{149}, and Asn^{240}) [130].

The N-terminal 300 amino acids are predicted to form six transmembrane α-helices, while the C-terminal cytoplasmic domain contains the binding sites for FAD and NADPH, shown experimentally through cross-linking studies and the observation that relipidated flavocytochrome alone can generate O_2^-. In addition, gp91^{phox} is responsible for complexing the two nonidentical heme groups of the NADPH oxidase via two histidine pairs. Hence gp91^{phox} contains all cofactors required for the electron transfer reaction which occurs in two steps. First, electrons are transferred from NADPH on to FAD and then to the heme group in the second step to reduce O_2 to O_2^- in a one-electron-transfer reaction. At present, no information is available on the three-dimensional structure of gp91^{phox} or fragments thereof, although a model for the structure of the cytoplasmic domain of gp91^{phox} has been suggested based on sequence homology with the FNR (ferredoxin-NADP reductase) family [131]. Significant insight into the topology of the cytochrome and the sites of interaction with other oxidase components has been gained through the use of a number of techniques, including epitope mapping or random sequence peptide phage analysis. Additionally, the study of cytochrome isolated from patients with X-linked CGD has contributed to our current understanding of its function [132].

Involvement of the gp91phox subunit in oxidative burst response by PMNs as well as Mϕs is not clear. Macrophages use a membrane-associated NADPH oxidase to generate an array of oxidizing intermediates. In some studies, it has been demonstrated that oxidants potently and efficiently inactivate matrilysin (MMP-7) by cross-linking adjacent tryptophan-glycine residues within the catalytic domain of the enzyme. These in vitro observations suggest that MMP inactivation can occur on or near phagocytes that produce both MMPs and reactive intermediates. In the absence of reactive intermediates, unrestrained proteolytic activity might lead to detrimental tissue damage. Indeed, inherited deficiency of gp91^{phox}, a phagocyte-specific component of the NADPH oxidase required for oxidant production, and targeted deletion of its mouse homologue result in granuloma formation and excessive tissue destruction [133]. Aberrant regulation of MMP activity may contribute to the damage that occurs when phagocytes are unable to generate oxidants. *gp91^{phox}* mutant mice were found to develop spontaneous, progressive emphysema, equal to that seen in smoke-exposed wild-type animals. Macrophages and neutrophils use membrane-associated NADPH oxidase to generate reactive oxygen intermediates. The initial product of the NADPH oxidase is superoxide, which is converted into oxidizing oxygen and nitrogen species. Because their characteristic end products have been detected in diseases ranging from atherosclerosis to neurodegenerative disorders, reactive oxygen and nitrogen intermediates are thought to contribute to inflammatory tissue injury. However, humans and animals deficient in phagocyte NADPH oxidase tend to form granulomas and to have excessive tissue destruction and an exuberant inflammatory response, raising the possibility that oxidants derived from white cells actually govern or suppress inflammation.

One potential mechanism whereby reactive oxygen species can influence inflammation and the associated tissue damage is by regulating the activity of MMPs. In addition to their ability to act on extracellular matrix, MMPs can affect inflammation by directly or indirectly regulating the activity of inflammatory mediators such as chemokines. Because reactive intermediates effectively inactivate MMPs in vitro, they provide an efficient mechanism for inhibiting unregulated catalysis by these extracellular proteinases, thereby preventing pathological destruction of tissue proteins and exuberant inflammation. Production of reactive intermediates by the phagocyte NADPH oxidase could confine MMP activity in space and time, permitting only bursts of pericellular proteolysis.

This study addresses for the first time the relationship between gp91phox and MMP in the development of T-cell-mediated acute allergic asthma in a mouse model using genetic knockout mice, gp91phox−/− which will be referred to as NOX−/− and MMP12NOX double knockout. The study shows that gp91phox most likely has a regulatory role in the onset and maintenance of the composite asthma phenotype in mouse and deletion of gp91phox may alter expression of costimulatory/co-inhibitory molecules B7.1 (increased) and B7.2 (decreased) and MHCII expression (increased) which may explain the mechanism by which macrophages despite increased migration to the inflammatory foci in vivo and increased migration in a chemotaxis chamber to MCP-1, and enhanced proliferation to syngeneic or allogeneic stimulus, fail to show oxidative burst response. MMP12 seems to be either the innocent by-stander, not contributing to the overall asthma phenotype, or has a synergistic (not additive) role in the process.

Materials and Methods

Mice

Both gp91$^{phox-/-}$ mice {Pollock, J. D., (1995) *Nat. Genet.* 9, 202} (Jackson Laboratories, Bar Harbor, ME) and mmp12$^{-/-}$ mice {Shipley, J. M (1996) *Proc. Natl. Acad. Sci. U. S. A.* 93, 3942–3946} were on a C57Bl/6 J background and had been outcrossed and then intercrossed for three generations to generate animals deficient in both genes. C57BL6 mice (Taconic) were used as the control group and are called wild type. In total the following number of animals was used in each group:

n	WT	NOX−/−	MMP12NOX−/−
Control (+Alum)	14	14	16
+OVA	16	15	14

Allergen Sensitization and Challenge

Mice were sensitized and later challenged with OVA (Pierce, Rockford, IL) as described previously (2). Mice were immunized with OVA (100 μg) complexed with aluminum sulfate in a 0.2-ml volume, administered by i.p. injection on day 0. On days 8 (250 μg of OVA) and on days 15, 18, and 21 (125 μg of OVA), mice were anesthetized briefly with inhalation of isoflurane in a standard anesthesia chamber and given OVA by intratracheal (i.t.) administration. Intratracheal challenges were done as described previously (Iwata A, JI, 2001). Mice were anesthetized and placed in a supine position on the board. The animal's tongue was extended with lined forceps, and 50 μl of OVA (in the required concentration) was placed at the back of its tongue. The control group received normal saline with aluminum sulfate by i.p. route on day 0 and 0.05 ml of 0.9 % saline by i.t. route on days 8, 15, 18, and 21.

Pulmonary Function Test

In vivo airway hyperresponsiveness to methacholine was measured 24 h after the last OVA challenge by both invasive and noninvasive plethysmography.

Invasive Plethysmography

On day 22, 24 h after the last intratracheal allergen (OVA) challenge, invasive pulmonary mechanics were measured in mice in response to methacholine in the same manner as previously described {Henderson WR, Jr., J Exp Med. 1996 184(4):1483–94} with the following modifications: (a) the thorax was not opened, (b) mice were ventilated with a tidal volume of 200 μl and respiratory rate of 120 breaths/min using a MiniVent Ventilator for Mice (Harvard Apparatus, Holliston, MA), (c) mice received aerosolized solutions of methacholine (0, 3.125, 6.25, 12.5, 25, 50, and 100 mg/ml in normal saline) via an AER 1021 nebulizer aerosol system (Buxco Electronics, Inc., Wilmington, NC) with 2.5–4 μm aerosol particle size generated by NEB0126 nebulizer head (Nektar Therapeutics, San Carlos, CA), and (d) a commercial plethysmography system (Model PLY4111 plethysmograph, MAX II amplifier and pressure transducer system, and Biosystem XA software, Buxco Electronics, Inc.) was used to determine R_L as calculated from measures of pressure and flow and expressed as $cmH_2O/ml/s$. Noninvasive plethysmography (expressed as Penh) was also assessed on day 22 in independent experiments.

Noninvasive Whole-Body Plethysmography

In conscious, free-moving, spontaneously breathing mice using whole-body plethysmography (model PLY 3211; Buxco Electronics, Sharon, CT) as previously described (Henderson WR 2005, J Allergy Clin Immunol) lung function was measured. Mice were challenged with aerosolized saline or increasing doses of methacholine (5, 20, and 40 mg/ml) generated by an ultrasonic nebulizer (DeVilbiss Health Care, Somerset, PA) for 2 min. The degree of bronchoconstriction was expressed as enhanced pause (P_{enh}), a calculated dimensionless value, which correlates with the measurement of airway resistance, impedance, and intrapleural pressure in the same mouse. P_{enh} readings were taken and averaged for 4 min after each nebulization challenge. Penh was calculated as follows: $P_{enh} = [(Te/Tr-1)X\ (PEF/PIF)]$, where T_e is expiration time, T_r is relaxation time, PEF is peak expiratory flow, and PIF is peak inspiratory flow × 0.67 coefficient. The time for the box pressure to change from a maximum to a user-defined percentage of the maximum represents the relaxation time. The T_r measurement begins at the maximum box pressure and ends at 40 %.

BALf

After pulmonary function testing, the mouse underwent exsanguination by intraorbital arterial bleeding and then BAL (0.4 ml three times) of both lungs. Total BAL fluid cells were counted from a 50-μl aliquot, and the remaining fluid was centrifuged at 200 g for 10 min at 4 °C, and the supernatants stored at –70 °C for assay of BAL cytokines later. The cell pellets were resuspended in FCS and smears were made on glass slides. The cells, after air-drying, were stained with Wright-Giemsa (Biochemical Sciences Inc, Swedesboro, NJ), and their differential count was taken under a light microscope at 40X magnification. Cell number refers to that obtained from lavage of both lungs/mouse.

Lung Parenchyma

Lung mincing and digestion was performed after lavage as described previously (Labarge S et al) with 100 u/ml collagenase for 1 h at 37 °C and filtered through a #60 sieve (Sigma). All numbers mentioned in this paper refer to cells obtained from one lung/mouse.

Lung Histology

Lungs of other animals of same group were fixed in 4 % paraformaldehyde overnight at 4 °C. The tissues were embedded in paraffin and cut into 5-μm sections. A minimum of 15 fields were examined by light microscopy. The intensity of cellular infiltration around pulmonary blood vessels was assessed by hematoxylin and eosin staining. Airway mucus was identified by staining with Alcian blue and periodic acid Schiff staining as described previously.

Fluorescein-Activated Cell Sorter (FACS) Analysis

Cells from hemolyzed peripheral blood (PB), bone marrow(BM), bronchoalveolar lavage (BAL), lung parenchyma (LP), spleen, mesenteric lymph nodes (MLN), cervical lymph nodes (CLN), axillary lymph nodes (LNX), and inguinal lymph nodes (LNI) were analyzed on a FACSCalibur (BD Immunocytometry Systems, San Jose, CA) by using the CellQuest program. Staining was performed by using antibodies conjugated to fluorescein isothiocyanate (FITC), phycoerythrin (PE), allophycocyanin (APC), peridinin-chlorophyll-protein (Per CP-Cy5.5), and Cy-chrome (PE-Cy5 and PE-Cy7). The following BD Pharmingen (San Diego, CA) antibodies were used for cell-surface staining: APC-conjugated CD45 (30 F-11), FITC-conjugated CD3(145-2C11), PE-Cy5-conjugated CD4 (RM4-5),PE-conjugated CD45RC (DNL-1.9), APC-conjugated CD8(53–6.7), PE-Cy5-conjugated β220 (RA3-6β2), FITC-conjugated IgM, PE-conjugated CD19 (ID3), PE-conjugated CD21(7G6), FITC-conjugated CD23 (B3B4), APC-conjugated GR-1(RB6-8C5), and PE-conjugated Mac1(M1/70). PE-Cy5-conjugated F4/80 (Cl:A3-1(F4/80)) was obtained from Serotec Ltd., Oxford, UK. PE-conjugated anti-α4 integrin (PS2) and anti-VCAM-1(M/K-2) were from Southern Biotechnology, Birmingham, Ala. Irrelevant isotype-matched antibodies were used as controls.

CFU-C Assay

To quantitate committed progenitors of all lineages, CFU-C assays were performed using methylcellulose semisolid media (Stemgenix, Amherst, N.Y.) supplemented with an additional 50 ng of stem cell factor (Peprotech, Rocky Hill, N.J.) per ml. Next, 50,000 cells from bone marrow, 500,000 cells from spleen, 0.01 million cells from lung and BAL, and 10 μl peripheral blood were plated on duplicate 35-mm culture dishes and then incubated at 37 °C in a 5 % CO_2-95 % air mixture in a humidified chamber for 7 days. Colonies generated by that time were counted by using a dissecting microscope, and all colony types (i.e., burst-forming units-erythroid [BFU-e], CFU-granulocyte-macrophage [CFU-GM], and CFU-mixed [CFU-GEMM]) were pooled and reported as total CFU-C. Total CFU-C per organ was calculated by extrapolating CFU-C against number of plated cells to the total number of cells in the organ.

ELISA for Cytokines

Th2 cytokines (IL-4 and IL-5) and TNFα and IFNγ in BAL and serum (previously frozen at –80^{0}C) were outsourced to LINCOplex Biomarker Immunoassays, Millipore. IL-13 was measured by Quantikine M kits from R&D Systems, Minneapolis, MN.

OVA-Specific IgE and IgG1 in Serum

Anti-mouse IgE (R35-72) and IgG1(A85-1) from BD Biosciences, San Diego, CA, were used for measuring OVA-specific IgE and IgG1 (in serum previously frozen at –70 °C) respectively by standard ELISA procedures as previously described (Henderson WR Jr. et al J Allergy Clin Immunol, 2005).

Chemotaxis Assay

Chemotaxis assay was performed with ten million macrophages pooled from 4 mice/experimental group. Macrophages were prepared by adhering BALf cells in high glucose medium for 2 h followed by detachment by mechanical scraping and resuspension in phenol red-free high glucose DMEM (Gibco) with 5 % FBS with 0.5 μg/ml Calcein-AM (1:2000 dilution) and incubation for 20 min at 37 °C. MCP-1 at dilutions ranging from 0.1 to 25 mM were used and 15 mM was taken to be the optimum dose. 96-well Neuroprobe CTX plates (Chemicon, Temecula, CA) were used. 29 μl MCP-1 (15 mM) was added as a single convex drop and the polycarbonate filter placed gently over it and incubated at 37 °C for 30 min. Cell suspension was added in designated slots over the filter membrane also in 29-μl volume. The chamber was incubated at 37 °C in humidified CO_2 incubator for 2 h. Excess cells were wiped off with Kimwipes at the end of the incubation period. Migrated cells were quantified by fluorescence (excitation at 488 nm, emission at 520 nm) using a Victor 3 V (Perkin Elmer laboratories) using a Wallac1420 software.

Mouse Model of Pulmonary Fibrosis

A single intratracheal dose of 0.075 U/ml of bleomycin in 40-μl saline was administered (d 0), and mice were sacrificed 14 and 21 days later. Mice C57Bl/6 were kept under ABSL-2 conditions approved by the IACUC of the University of Washington and monitored daily. They were housed under specific pathogen-free condition and were given food and water ad libitum. They were sacrificed on day 14. 1 week after bleomycin administration, mice developed marked interstitial and alveolar fibrosis, detected in lung sections by Masson's trichrome stain. Analysis of cell populations by enzymatic digestion by collagenase type IV followed by cell counting in Coulter counter and subsets identified and quantified by FCM and total and differential count of H&E stained cytospin smears of single-cell suspensions show loss of type II and type I alveolar epithelial cells and influx of macrophages. AEI and II were isolated following standard protocol.

Real-Time-PCR Analysis

For real-time-PCR analysis of mRNA expression of particular genes in differentiating human ESC as well as in the lungs of recipient animals in transplantation experiments, total RNA was extracted from cells (<500/sample) by PicoPure RNA isolation kit from Arcturus, Mountainview, CA, and those from lungs (kept in RNAlater (Ambion) at -80 °C), by RNA extraction kit (RNeasy) from Qiagen and cDNA made from it using superscript III system from Invitrogen. The PCR reaction solution contained 0.5 μg of total RNA, 6-mM magnesium chloride, and 0.5 μM of each primer (primer oligo sequences are in Table 3.8). Other components in the reverse transcriptase-PCR master mix included buffer, enzyme, SYBR Green I, and deoxyribonucleotide triphosphate. For reverse transcription, the 20 μL of reaction capillaries was incubated at 50 °C for 2 min followed by a denaturation at 95 °C for

Table 3.8 List of mouse primers for real-time PCR

Cell marker	Gene	Forward primer	Reverse primer
Housekeeping genes	GAPDH	CGTCCCGTAGACAAAATGGT	TCAATGAAGGGGTCGTTGAT
	β-actin	GTGGGCCGCTCTAGGCACCAA	CTCTTTGATGTCACGCACGATTTC
Cytokine genes	IFN-γ	GCGTCATTGAATCACACCTG	TGAGCTCATTGAATGCTTGG
	IL-1α	TCAAGATGGCCAAAGTTCCT	TGCAAGTCTCATGAAGTGAGC
	IL-1β	TGAAGCAGCTATGGCAACTG	GGGTCCGTCAACTTCAAAGA
	IL-2	AACCTGAAACTCCCCAGGAT	CGCAGAGGTCCAAGTTCATC
	IL-3	CCGTTTAACCAGAACGTTGAA	CCACGAATTTGGACAGGTTT
	IL-4	GGCATTTTGAACGAGGTCAC	AAATATGCGAAGCACCTTGG
	IL-5	ATGGAGATTCCCATGAGCAC	AGCCCCTGAAAGATTTCTCC
	IL-6	AACGATGATGCACTTGCAGA	GGTACTCCAGAAGACCAGAGGA
	IL-10	TGAATTCCCTGGGTGAGAAG	TGGCCTTGTAGACACCTTGG
	IL-12β	ATCGTTTTGCTGGTGTCTCC	CATCTTCTTCAGGCGTGTCA
	IL-13	CCTCTGACCCTTAAGGAGCTT	ATGTTGGTCAGGGAATCCAG
	MCP-3	TCTGTGCCTGCTGCTCATAG	CTTTGGAGTTGGGGTTTTCA
Growth factor genes	TGFβ2	GGAGGTTTATAAAATCGACATGC	GGCATATGTAGAGGTGCCATC
	VEGFa	TACCTCCACCATGCCAAG	TGGTAGACATCCATGAACTTGA
	VEGFb	GGCTTAGAGCTCAACCCAGA	TGGAAAGCAGCTTGTCACTTT
	VEGFc	GGGAAGAAGTTCCACCATCA	TCGCACACGGTCTTCTGTAA

10 min. Polymerase chain reaction was started off by an initial denaturation at 95 °C for 15 s and then annealing at 60 °C 1 min, repeat 45 cycles. Finally, a melting curve analysis was performed by following the final cycle with incubation at 95 °C for 15 s, at 60 °C for 15 s, and then 95 °C for 15 s. Negative control samples for the reverse transcriptase-PCR analysis, which contained all reaction components except RNA, were performed simultaneously to determine when the nonspecific exponential amplification cycle number was reached. Forward and reverse primers are as in Table 3.8 and were synthesized by the University of Washington Biochemistry services using the Primer Express software. PCR was carried out using the comparative Ct method (Applied Biosystems software) with SYBR Green PCR core reagents (Applied Biosystems) and analyzed using Applied Biosystems 7900HT Real-Time-PCR System software SDS 2.2.1.

Statistical Analysis

Statistical differences among samples were tested by Student's *t*-test. *P* value less than 0.05 was considered statistically significant.

Results

Effect of gp91^{phox} Deletion on Development of Composite Asthma Phenotype

Composite asthma phenotype developed completely and more aggressively in both gp91phox–/– and MMP12-gp91phox double knockout mice post-OVA treatment as described in Fig. 3.24. Inflammation in terms of total inflammatory migration (measured by FACS of surface marker expression and differential count of H&E stained cytospin smears) was increased (Table 3.9). In bone marrow, NOX–/– post-OVA has 1.4-fold more cells, peripheral blood has 1.3-fold more cells, spleen had 1.3-fold more cells, lung parenchyma had 1.8-fold more cells, and BALf had twofold more cells compared to post-OVA WT. Of note, 2.4-fold more PMNs, 1.96-fold more B lymphocytes, and fivefold more eosinophils in post-OVA NOX–/– compared to post-OVA WT BALf were found. Overall systemic response, inflammatory recruitment from blood to inflammation in the lung is more (Table 3.10).

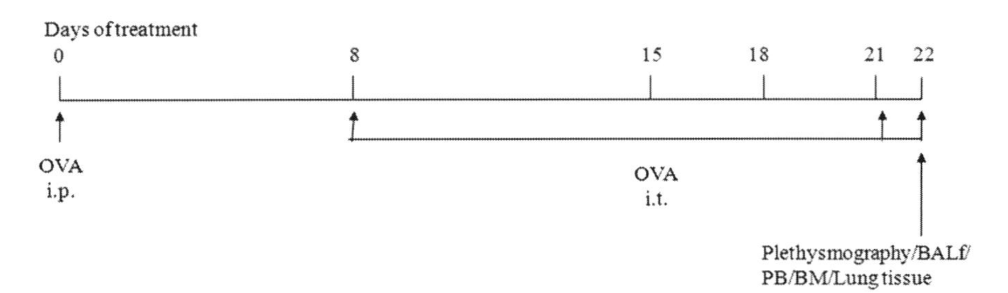

Fig. 3.24 *Study design to generate acute allergic asthma phenotype in mice*: mice were sensitized with OVA (100 μg) complexed with aluminum sulfate in a 0.2-ml volume, administered by i.p. injection on day 0 and later challenged with OVA on day 8 (250 μg of OVA) and on days 15, 18, and 21 (125 μg of OVA) (Pierce, Rockford, IL) by intratracheal instillation. Mice were immunized with OVA (100 μg) complexed with aluminum sulfate in a 0.2-ml volume, administered by i.p. injection on day 0. The control group received normal saline with aluminum sulfate by i.p. route on day 0 and 0.05 ml of 0.9 % saline by i.t. route on days 8, 15, 18, and 21.They were sacrificed on day 22. The abbreviations used are as follows: *i.p.* intraperitoneal, *i.t.* intratracheal, *BAL* bronchoalveolar lavage, *PB* peripheral blood

Pulmonary Function Test

Pulmonary function test measured by both non-invasive plethysmography and invasive plethysmography (Fig. 3.25a–c) shows 2.5-fold increase in Penh values compared to WT post-OVA at 100 mg/ml dose of methacholine. 1.89- and 1.35-fold increase was found in gp91phox−/− and DKO respectively at 100 mg/ml dose of methacholine compared to WT post-OVA by invasive plethysmography of anesthetized mice.

Cellularity in Bone Marrow, Blood, Lungs, and Airways

Cellularity was increased in post-OVA mice compared to saline-treated mice. Figure 3.26 shows 1.8-fold and 1.7-fold increase in number of cells in lung parenchyma and BALf and 1.4-fold in spleen in both KO mice than WT. Recruitment index (Table 3.10) was increased in B cells, monocytes, PMNs, eosinophils, and basophils in both KO mice post-OVA compared to post-OVA WT.

Progenitors in Bone Marrow, Blood, Lungs, and Airways

Table 3.11 and Fig. 3.27 show decrease in number of colony-forming units in post-OVA bone marrow and lungs of both KO mice compared to similarly treated WT. Progenitor number was however somewhat upregulated in both KO groups post-OVA compared to Post-OVA WT.

Inflammation in the Lungs

Histopathology showed a marked increase in peribronchiolar and perivascular infiltration of inflammatory cells post-OVA in both KO mice compared to control (Fig. 3.28a–c). The photomicrographs show a marked increase in both number of inflammatory cells and mucus secretion in airways.

Airway Goblet Cell Metaplasia

1.27-fold and 1.22-fold increase in percent metaplastic goblet cells (Fig. 3.29) was measured by counting under light microscope, the blue mucus laden cells around airways versus the pink squamous cells which do not show metaplasia.

Th2 Cytokine Release in Airway

Cytokine concentrations present in the BALf were measured by ELISA. Figure 3.30 shows 2.75-fold increase in IL-13 level in gp91phox−/− mice post-OVA compared to WT post-OVA. IL-4 was increased. All other Th2 cytokines showed values similar to post-OVA WT BALf. Table 3.12 shows that actual mRNA upregulation was 1.4-fold for IL-4 gene and 1.9-fold for IL-13 genes which are Th2 specific. There was also upregulation in IL-1α, IL-10, and IL-12α, the significance of which is not clear at this point. Cells secreting these are T cells, MACS, and some epithelial cells. Overall, IL-4: NOX−/− post-OVA has 1.2-fold more protein

Table 3.9 Number of cells (×106) of leucocyte subsets in peripheral blood, lung parenchyma, and bronchoalveolar lavage fluid (BALf). Leukocyte subsets (×106) in blood, lungs, and BAL fluid post-OVA treatment

Blood								
	Total	**Lymphocytes**						
	WBCs	**CD3+**	**β220+**	**Monocytes**	**Neutrophils**	**Eosinophils**	**Basophils**	
WA	11.9	5.54	3.08	0.38	3.648	0.223	0.066	
	±2.4	±1.17	±1.05	±0.185	±0.184	±0.014	±0.001	
WO	24.72	7.31	5.93	1.62	7.97	0.74	1.05	
	±6.41	±1.31	±1.17	±0.35	±1.14	±0.36	±0.02	
NOXA	14.2	5.7	2.99	0.77	4.56	0.133	0.028	
	±3.012	±1.865	±1.033	±0.053	±1.074	±0.043	±0.002	
NOXO	25.625	6.186	6.742	2.424	8.741	**1.061***	0.556	
	±4.063	±1.076	±0.964	±1.076	±0.064	±0.004	±0.003	
DKOA	11.7	4.69	2.467	0.63	3.74	0.106	0.02	
	±2.76	±1.06	±0.96	±0.12	±1.27	±0.07	±0.001	
DKOO	27.56	6.61	7.27	2.6	9.39	**1.14***	0.59	
	±2.79	±1.49	±2.01	±0.47	±2.44	±0.46	±0.14	
Lungs								
	Total	**Lymphocytes**					**Basophils/**	
	WBCs	**CD3+**	**β220+**	**Monocytes**	**Neutrophils**	**Eosinophils**	**Mast cells**	**Mϕ**
WA	1.86	0.179	0.035	0.826	0.319	0	0	0.497
	±0.543	±0.021	±0.002	±0.054	±0.054	±0.015		
WO	12.35	4.87	0.09	0.37	1.47	2.22	0.046	3.09
	±3.96	±1.05	±0.01	±0.09	±0.43	±0.76	±0.012	±1.08
NOXA	**5.49#**	**0.446#**	**0.113#**	**2.593#**	**0.892#**	0	0	**1.489#**
	±1.064	±0.002	±0.001	±0.76	±0.011	±0.016		
NOXO	**22.6***	4.084	**1.29***	**3.171***	**4.337***	**4.102***	**0.089***	5.682
	±7.544	±1.075	±0.643	±1.075	±1.074	±1.066	±0.002	±0.064
DKOA	**3.78#**	0.3	0.07	**1.78#**	0.61	0.001	0	**1.02#**
	±1.07	±0.14	±0.01	±0.54	±0.02	±0.0001	±0.29	
DKOO	**21.66***	3.89	**0.43***	**3.03***	**4.16***	**3.93***	**0.08***	**5.44***
	±408	±1.06	±0.16	±0.89	±1.33	±1.13	±0.002	±1.26
BAL fluid (2 lungs)								
WA	0.8	0	0	0.013	0	0	0	0.787
	±0.013	±0.004	±0.011					
WO	9.46	1.64	1.54	1.26	0.86	4.26	0.04	0.56
	±0.89	±0.21	±0.021	±0.03	±0.01	±0.07	±0.002	±0.004
NOXA	0.63	0	0	0.014	0	0	0	0.616
	±0.018	±0.002	±0.003					
NOXO	**16***	2.04	1.213	2.08	**2.427***	**6.741***	0.051	**1.568***
	±3.159	±0.754	±0.003	±0.074	±1.066	±2.076	±0.001	±0.72
DKOA	0.865±0.01	0	0	0.025±0.002	0	0	0	0.84±0.18
DKOO	**15.89***	2±0.86	1.22±0.66	2.05±0.75	**2.415**	**6.67***	0.05	**1.55***
	±4.03	±1.01*	±1.14	±0.003	±0.63			

Total white blood cells were counted by a Beckman Coulter particle counter and subsets quantified using specific antibody staining in FACS and corroborated by morphological assessment by light microscopy and also by hemavet using the specifics of mouse. The table shows data averaged from three independent experiments ± SEM

Abbreviations used are as follows: *WA* wild type +alum, *WO* wild-type + OVA, *NOXA* gp91phox−/− + alum, *NOXO* gp91phox−/− + OVA, *DKOA* MMP12-gp91phox double knockout + alum, *DKOO* MMP12-gp91phox double knockout + OVA. (n=5/group)

*denotes p value < 0.05 compared to values from post-OVA wild-type mice

$^\#$denotes p value < 0.05 compared to alum-treated wild type

Table 3.10 Recruitment index: Inflammatory hematopoietic cells found in the lungs and BALf expressed as a fraction of corresponding cells in circulation. The number of cells ($\times 10^6$) in the lungs and BALf was the numerator, and the number of cells ($\times 10^6$) in circulating blood was the denominator in the fraction calculated to show recruitment index or inflammatory cells recruited from circulation to lung parenchyma and airways in response to the OVA allergic challenge

		T cells	B cells	Monocytes	Neutrophils	Eosinophils	Basophils
Lungs	WO	0.6	0.015	0.22	0.18	3	0.04
	NOXO	0.6	0.19	1.3	0.49	3.8	0.16
	DKOO	0.58	0.05	1.1	0.44	3.4	0.13
BALf	WO	0.22	0.25	0.77	0.107	5.75	0.03
	NOXO	0.32	0.17	0.85	0.27	6.3	0.09
	DKOO	0.51	0.16	0.78	0.26	5.8	0.08

and twofold more mRNA; IL-5: NOX−/− post OVA has twofold more protein and 2.8-fold more mRNA; IL-13: NOX−/− post OVA has threefold more protein and 5.6-fold more mRNA. Therefore, both by protein concentration and mRNA expression, Th2 cytokines show manifold increase in gp91phox−/− post-OVA compared to WT.

Chemokine and Growth Factor Gene Expression

MCP-3 is a known macrophage chemotactic protein. Its gene expression was found to be upregulated manifolds both in the saline-treated control lung and post-OVA. The reason for this may be that in the absence of gp91phox, there is spontaneous upregulation of the chemokine gene. VEGFb was upregulated by 2.2-fold. Surprisingly, there was downregulation of TGFβ which might indicate that in keeping with decreased iNOS expression and the consequent shift in macrophage phenotype to M1 (killer) from M2 (healer), TGFβ expression was also downregulated corroborating once again that gp91phox may have a regulatory role in the development of Th2 phenotype and deletion of that disrupts the control or moderating effect involving cross talk between T cells and the downstream phagocytes which need NADPH enzyme for the respiratory burst response and proliferation. Increased chemotaxis to MCP-1 may be explained by the increased expression of MCP-3. Upregulation of MMP12 in gp91phox−/− both before and after OVA may indicate a compensatory mechanism in the regulation of Th2 response (Table 3.13).

Expression of Rho Kinase and MMPs Associated with Inflammation

Since this is an inflammatory response to the allergen, we hypothesized that other proinflammatory kinases and proteases may also be intimately involved in the pathway. Figure 3.31 shows RT-PCR analysis of gene expression of Rho kinase, traditionally known to have anti-inflammatory functions, to be downregulated, and so are MMP7, 10, and 28. The metalloproteases are also known regulators of inflammation, and their downregulation again corroborates that deletion of gp91phox may have disrupted a control mechanism on the development of Th2-mediated inflammation in the lung.

Functionality of B Cells

B-cell function was tested by measuring OVA-specific IgE and IgG1 in post-OVA WT and the two KO mice. Figure 3.32 shows that the values are comparable.

Functionality of T Cells

Proliferation of MACS-purified (>86–92 %) CD4+ and CD8+ splenocytes by MTT incorporation assay and OD measurement at 545 nm of anti-CD3/CD28 (0.01–1 ug/ml)-induced proliferation of CD4+ shows a 8.4-folds increase in post-OVA WT compared to a 7.4-fold increase in both KO mice. In CD8+ while post-OVA WT increased by 6.4-fold, the KO mice showed 7.9-fold increase compared to the corresponding saline-treated

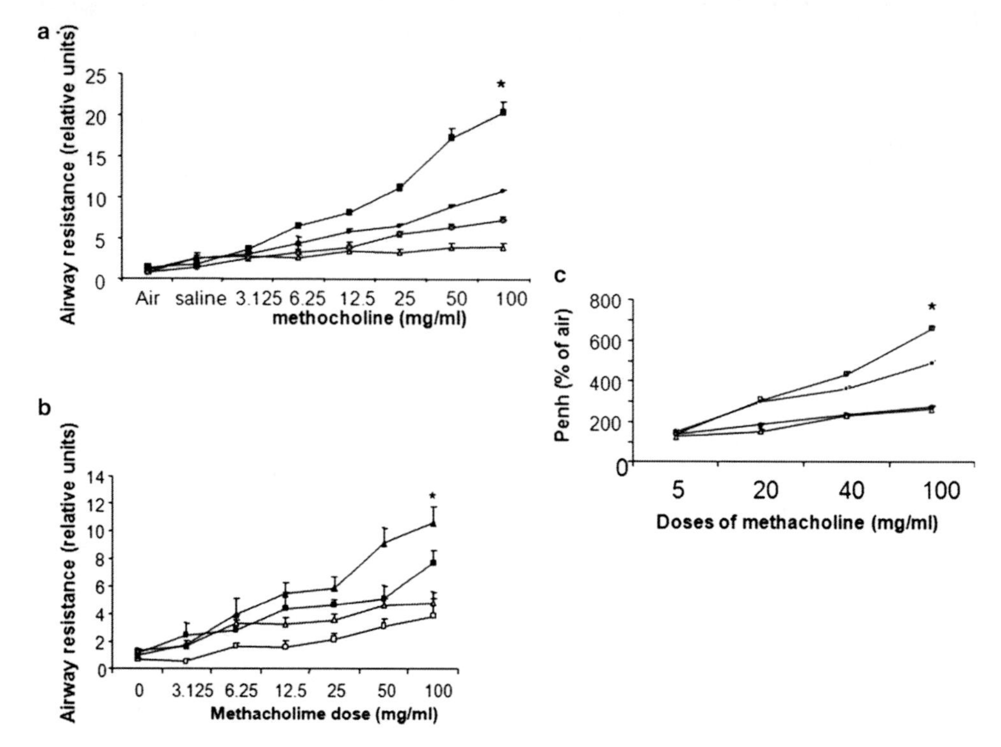

Fig. 3.25 *Increase in total cell number in the lung and BALf of knockout mice compared to post-OVA WT*. Cell number was counted using a Z1 particle counter from Beckman Coulter. Bone marrow cells (BM) were flushed out of two femurs, blood (PB) was obtained by infra-orbital bleeding and extrapolated to a 2-ml volume as the total volume of blood in a 20-gm mouse, from the volume of blood actually obtained, perfused lung was minced and digested with collagenase IV and single-cell suspension made of both lungs, and bronchoalveolar lavage fluid (BALf) was obtained from both lungs, and the cell numbers were counted. The data shown have been derived from three independent experiments and expressed as mean values ± SEM. *denotes p value < 0.01 compared to post-OVA wild-type values. Abbreviations used are as follows: *WT* wild type, *NOX* gp91phox−/−, *DKO* gp91phox-MMP12 double knockout, *WA* WT+alum, *WO* WT+OVA, *NOXA* gp91phox−/−+alum, *NOXO* gp91phox−/−+OVA, *DKOA* gp91phox-MMP12 double knockout+alum, and *DKOO* gp91phox-MMP12 double knockout+OVA. *Pulmonary function testing*. (**a**). Airway resistance in WT versus gp91phox−/−. Symbols

—△— WA
—●— WO
—⊠— NOXA

used are as follows: —■— NOXO. (**b**). Airway resistance obtained by invasive plethysmography shows increased values in both gp91phox−/− and gp91phoxMM12 double knockout mice compared to wild-type post-OVA. Symbols

—□— WA
—■— WO
—△— DKOA

used are as follows: —▲— DKO . On day 22, 24 h after the last intratracheal allergen (OVA) challenge, invasive pulmonary mechanics were measured using a commercial plethysmography system (Model PLY4111 plethysmograph, MAX II amplifier and pressure transducer system, and Biosystem XA software, Buxco Electronics, Inc.), mice received aerosolized solutions of methacholine (0, 3.125, 6.25, 12.5, 25, 50, and 100 mg/ml in normal saline) via an AER 1021 nebulizer aerosol system (Buxco Electronics, Inc., Wilmington, NC) with 2.5–4 micron aerosol particle size generated by NEB0126 nebulizer head (Nektar Therapeutics, San Carlos, CA), and RL as calculated from measures of pressure and flow and expressed as cmH2O/ml/s was determined. Figures show airway resistance (RL) values ± SEM of data obtained from three independent experiments ($n = 5$/group). (**c**). Noninvasive plethysmography (expressed as Penh) was also assessed on day 22 in independent experiments with gp91phox−/− only, and this correlated with the airway resistance values obtained by invasive plethysmography. Symbols used are

—◆— WA
—▲— NOX
-·- WO

as follows: —■— NOX. For noninvasive plethysmography, mice were challenged with increasing doses of aerosolized methacholine (0, 5, and 20, 40, and 100 mg/ml in normal saline) generated by an ultrasonic nebulizer (DeVilbiss Health Care, Inc., Somerset, PA) for 2 min., and the degree of bronchoconstriction was expressed as enhanced pause (Penh), a calculated dimensionless value that correlates with measurement of airway resistance, impedance, and intrapleural pressure. Penh is primarily independent of FRC, tidal volume, and respiratory rate since the ratio of measurements is obtained during the same breath and has a strong correlation with both airway resistances (Raw) measured directly in anesthetized, tracheotomized, and mechanically ventilated mice. *denotes p value < 0.01 compared to post-OVA wild-type values

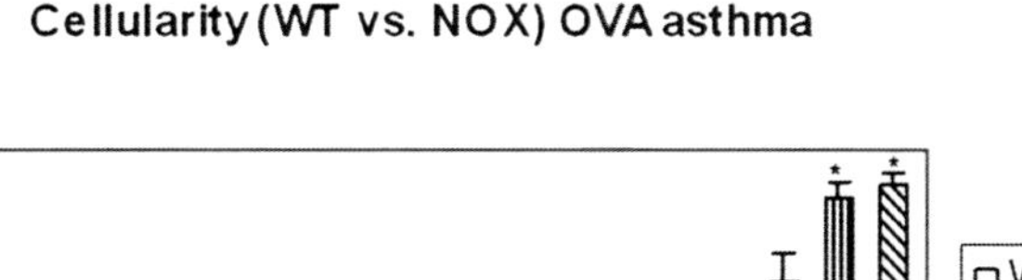

Fig. 3.26 *Decrease in clonogenic potential in bone marrow and lung of both KO mice post-OVA*. In gp91phox−/− progenitors in BM decrease by 28.8 %, while in double knockout, it decreases by 27.9 % compared to the values in WT post-OVA. In the lungs, the decrease in progenitor number was 1.8-fold and 1.5-fold respectively. Results shown are pooled from three independent experiments ($n = 5$/group) ± SEM. Abbreviations used are as follows: *BM* bone marrow, *PB* peripheral blood, *LP* lung parenchyma, *BALf* bronchoalveolar lavage fluid. The total number of progenitors was extrapolated to the total number cells obtained from each tissue. 10 ml heparinized whole peripheral blood (PB) obtained by infra-orbital bleeding of anesthetized mouse, 50,000 bone marrow cells (BMC) flushed out of the femurs, one million lung parenchyma (LP) cells digested by collagenase IV, one million cells from bronchoalveolar lavage fluid (BALf) and 500,000 cells from spleen were plated in 2 ml semi solid methyl cellulose medium and cultured for 7 days for PB, BM, and spleen and for 14 days for LP and BALf. Colony-forming units (CFU) were counted manually on a Leica DMIL and an Olympus sZX12 inverted microscope. Results shown are pooled from three independent experiments ($n = 5$/group) ± SEM. *denotes p value < 0.01 compared to post-OVA wild-type values. Abbreviations used are as follows: *WT* wild type, *NOX* gp91phox−/−, *DKO* gp91phox-MMP12 double knockout, *WA* WT+alum, *WO* WT+OVA, *NOXA* gp91phox−/−+alum, *NOXO* gp91phox−/−+OVA, *DKOA* gp91phox-MMP12 double knockout+alum, and *DKOO* gp91phox-MMP12 double knockout+OVA

Table 3.11 Decrease in clonogenic potential in bone marrow and lung of both KO mice post-OVA. Clonogenic potential of progenitors

	PB/2 ml	BM/2femurs	LP/2lungs	BAL/2lungs	Spleen
WA	5371.61 ± 188.6942	17034.5 ± 158.8784	241.93 ± 115.8199	3.15 ± 16.32993	75858.9 ± 523.0583
WO	16915.61 ± 219.1485	57161.5 ± 161.9139	11225.49 ± 155.0757	5573.87 ± 41.63332	188275.6 ± 361.9139
NOXA	2343 ± 244.3471	8293.32 ± 197.1253	604.12 ± 125.8199	59.926 ± 19.14854	41774.25 ± 75.49834
NOXO	9403.35 ± 382.2598	40677.7 ± 220.4159	6299.337 ± 195.9166	5921.744 ± 41.63332	219677.5 ± 426.8858
DKOA	4591.678 ± 256.6731	11783.19 ± 284.5762	562.91 ± 183.7863	32.932 ± 45.68423	65974.3 ± 344.3138
DKOO	9262.11 ± 342.6876	41278.3 ± 361.6896	7103.16 ± 201.5763	6012.22 ± 59.6876	197822.7 ± 388.6876

In gp91phox−/− progenitors in BM decrease by 28.8 % while in double knockout, it decreases by 27.9 % compared to the values in WT post-OVA. In the lung, the decrease in progenitor number was 1.8-fold and 1.5-fold, respectively. Results shown are pooled from three independent experiments (n = 5/group) ± SEM

Abbreviations used are as follows: *BM* bone marrow, *PB* peripheral blood, *LP* lung parenchyma, *BALf* bronchoalveolar lavage fluid

The total number of progenitors was extrapolated to the total number cells obtained from each tissue. Underlined numbers denote p value < 0.01 compared to post-OVA WT

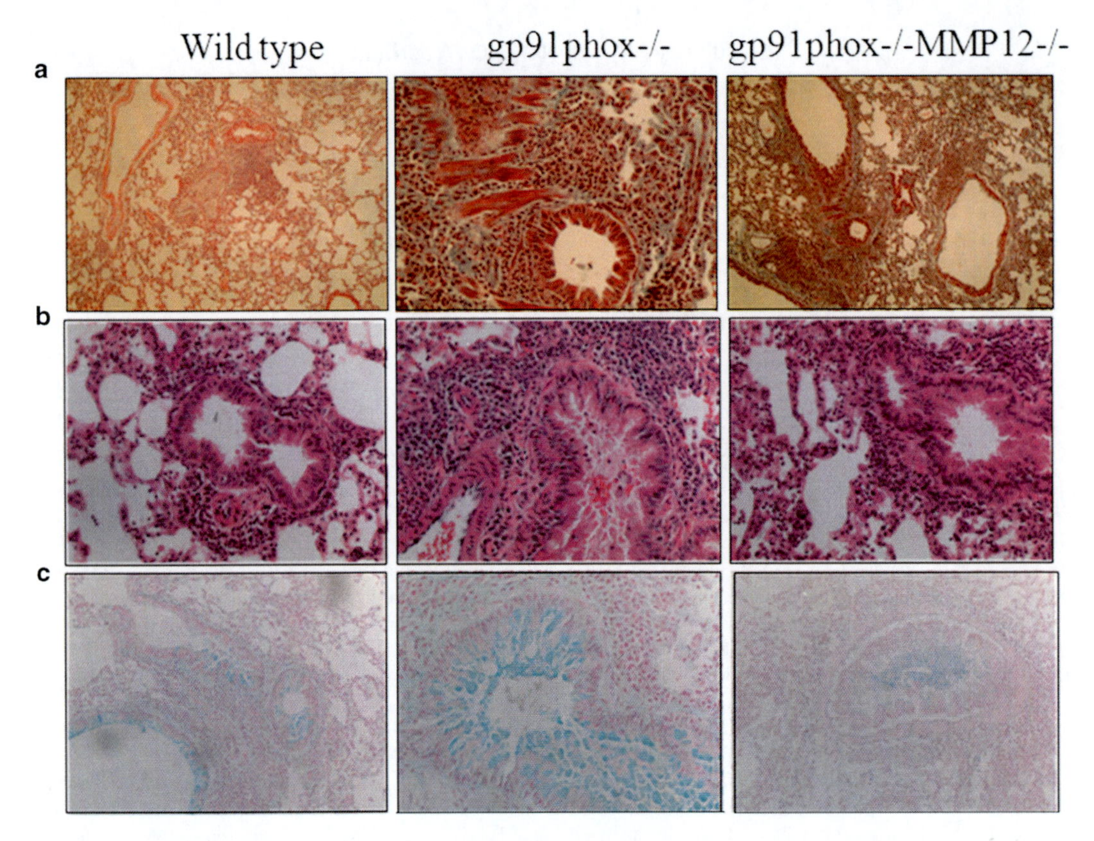

Fig. 3.27 *Both knockout mice show increased inflammation and mucus secretion post-OVA compared to WT.* Paraffin sections of lungs of OVA-treated mice were stained with (**a**), hematoxylin and eosin (10×), (**b**) with Masson's trichrome stain (20×), and (**c**). Alcian blue counterstained with eosin (10×). Sections were viewed with a BX4l Olympus microscope and photomicrographs taken with a Nikon Spot digital camera. Inflammatory cell migration was more pronounced around airways in the OVA-treated knockout mice compared to control

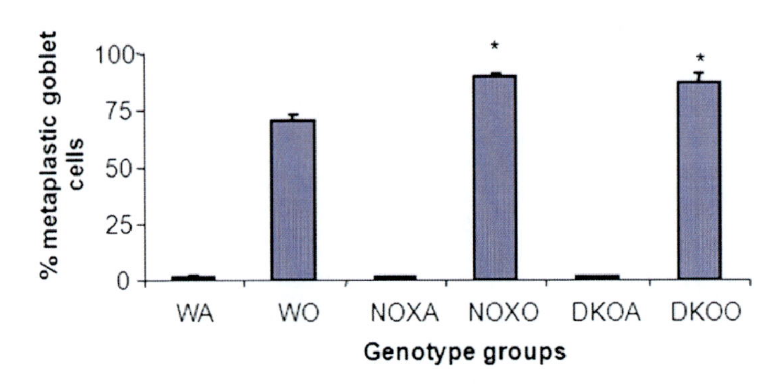

Fig. 3.28 *Increase in percent Alcian blue positive cells (goblet cell metaplasia) in gp91phox−/− and gp91phox-MMp12 double knockout (DKO) post-OVA compared to wild-type (WT) post-OVA.* Metaplastic goblet cells were counted as the blue cells stained positive by Alcian blue. Percent metaplastic goblet cells are calculated from the total number of cells counted around each airway. Abbreviations used are as follows: *WT* wild type, *NOX* gp91phox−/−, *DKO* gp91phox-MMP12 double knockout, *WA* WT+alum, *WO* WT+OVA, *NOXA* gp91phox−/−+alum, *NOXO* gp91phox−/−+OVA, *DKOA* gp91phox-MMP12 double knockout+alum, and *DKOO* gp91phox-MMP12 double knockout+OVA. The results shown are pooled from three independent experiments. Counts were taken and averaged over 10 high power fields. Slides were counted on a Spencer AO light microscope at 40× magnification. Results are expressed as percent ± SEM. $n = 5$ animals/group. *denotes p value < 0.01 compared to post-OVA wild-type values

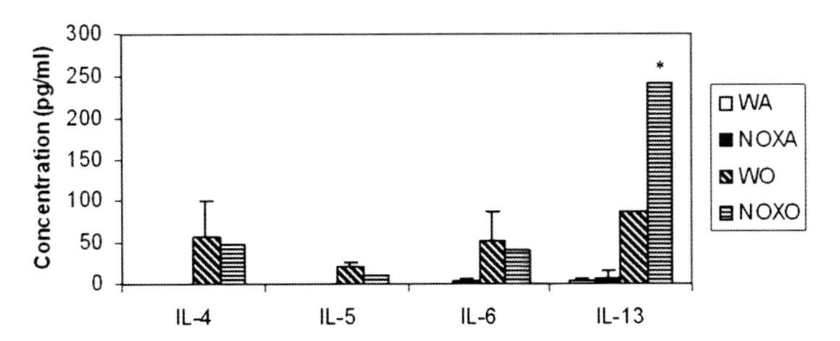

Fig. 3.29 *Cytokine concentration in BALf.* The concentration of cytokines in BALf was measured by outsourcing to Linco by multiplexing technique in a luminometer. Data expressed here are mean ± SEM. $n = 5$/group. While all other Th2 cytokine levels were comparable to WT + OVA, IL-13 concentration was increased 2.7-fold over post-OVA WT values. Abbreviations used are as follows: *WT* wild type, *NOX* gp91phox−/−, *DKO* gp91phox-MMP12 double knockout, *WA* WT+alum, *WO* WT+OVA, *NOXA* gp91phox−/−+alum, *NOXO* gp91phox−/−+OVA

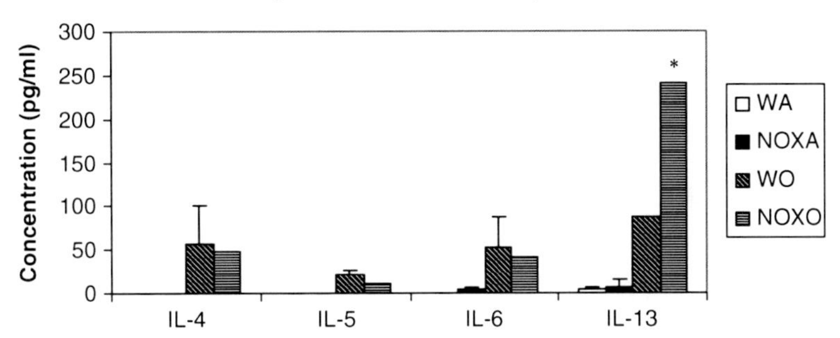

Fig. 3.30 Cytokine levels in WT, NOX-/- single knockout and NOX-/-MMP-12-/- double knockout mice respectively. Sandwich ELISA was performed with purified and biotinylated secondary antibodies and reading of OD taken at 405 nm and concentration expressed in pg/ml from a standard curve of increasing concentration. P denotes p value <0.01 was considered significant relative to control values and error bars provide information about mean ± SEM of three independent experiments of sample analyses in triplicate wells.

Table 3.12 Increase in gene expression of IL-1β, IL-10, IL-12 β, and IL-13. Real-time PCR analysis shows expression of mRNA for the particular genes as calculated by relative index of Ct values normalized to GAPDH by real-time PCR. PCR was carried out using the comparative Ct method (Applied Biosystems software) with SYBR Green PCR core reagents (Applied Biosystems) and analyzed using Applied Biosystems 7900HT Real-Time PCR System software SDS 2.2.1. All primers used were specific to mouse

	IFN-γ	IL-1α	IL-1 β	IL-2	IL-4	IL-5	IL-6	IL-10	IL-12β	IL-13
CWT	4.67 ± 1.70	71.27 ± 6.47	3.79 ± 0.32	2.57 ± 0.02	44.26 ± 12.70	27.15 ± 1.70	3.59 ± 1.65	4.21 ± 2.60	7.27	4.72
WTO	4.96 ± 1.85	411.67 ± 2.46	21.17 ± 8.11	65.56 ± 13.28	65.42 ± 13.58	103.45 ± 30.68	20.04 ± 9.80	23.61 ± 3.66	21421.95	45.15
CKO	1.18 ± 0.27	62.81 ± 0.68	6.20 ± 1.58	2.65 ± 1.61	33.72 ± 19.35	43.12 ± 0.90	3.74 ± 0.96	1.76 ± 0.65	29.23	4.68
KOO	2.06 ± 0.60	6037.51 ± 5.50	33.65 ± 1.74	8.63 ± 1.72	92.78 ± 1.02	147.74 ± 1.42	20.58 ± 8.28	134.96 ± 3.43	306948.33	90.59

The graphs in *denotes p value < 0.01 compared to bleo-treated untransplanted group. Mean denotes the average of 2 independent experiments ($n = 4$/group)

Abbreviations used are as follows: *cWT* saline-treated wild type, *WTO* wild-type post-OVA treatment, *CKO* saline-treated gp91phox knockout, and *KOO* CKO=OVA-treated gp91phox knockout

Underlined numbers denote mRNA expression in relative units normalized to mouse GAPDH that have p value < 0.05 compared to wild type post-OVA

Table 3.13 Increase in MCP-3 and VEGFb and decrease in TGFβ, VEGFa, and VEGFc in post-OVA gp91phox−/− lung. Real-time PCR data shows expression of mRNA for the particular genes as calculated by relative index of Ct values normalized to GAPDH by real-time PCR. PCR was carried out using the comparative Ct method (Applied Biosystems software) with SYBR Green PCRcore reagents (Applied Biosystems) and analyzed using Applied Biosystems 7900HT Real-Time PCR System software SDS 2.2.1. All primers used were specific to mouse

	MCP-3	TGFβ2	VEGFa	VEGFb	VEGFc
CWT	4.63 ± 1.63	18.03 ±15.34	1973.06 ± 1549.62	176.39 ± 2.24	2283.58 ± 1671.63
WTO	7.14 ± 1.94	44.67 ± 22.84	4260.86 ± 1982.14	334.59 ± 3.56	4500.67 ± 1827.83
CKO	20901.40 ± 6361.97	32.88 ± 9.71	34.40 ± 13.79	106.73 ± 80.67	15.11 ± 0.97
KOO	166575.87 ± 119300.10	28.30 ± 23.65	297.97 ± 234.29	744.17 ± 61.60	35.15 ± 11.41

The underlined numbers denote p value < 0.01 compared to OVA-treated wild-type group
Abbreviations used are as follows: *cWT* saline-treated wild type, *WTO* wild-type post-OVA treatment, *CKO* saline-treated gp91phox knockout, and *KOO* CKO=OVA-treated gp91phox knockout

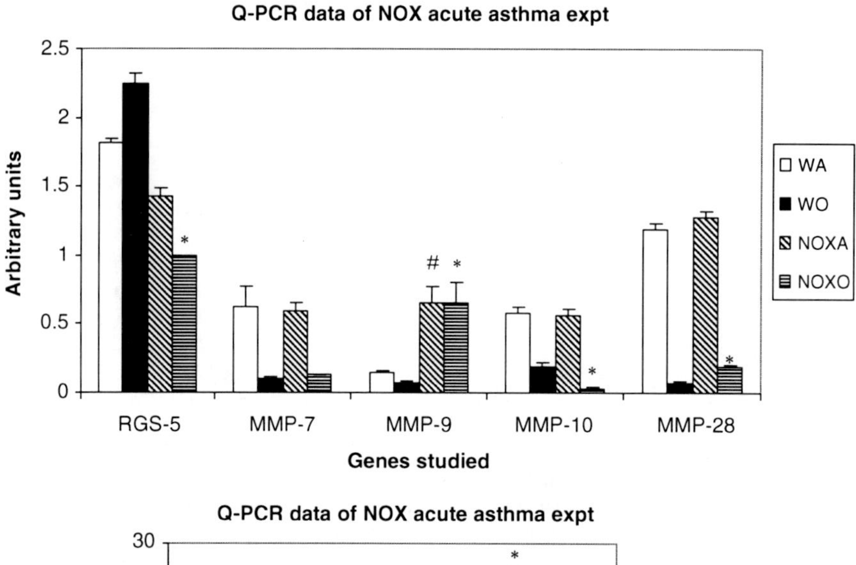

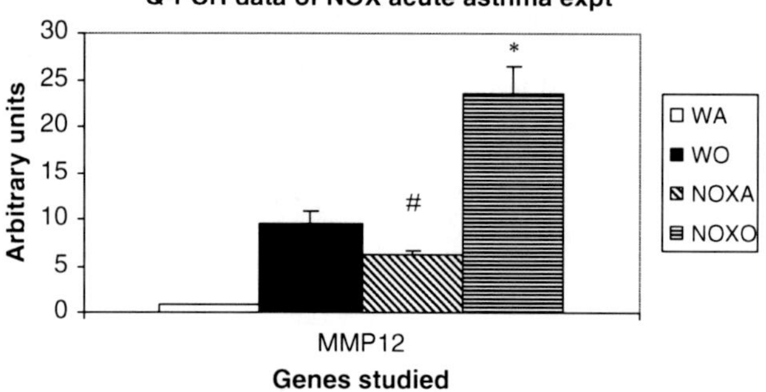

Fig. 3.31 *Downregulation Rho kinase RGS-5 and MMP10 but upregulation of MMP9 and MMp28 mRNA post-OVA in KO mice compared to WT*. Real-time-PCR analysis was used to quantitate the expression of mRNA for the particular genes as calculated by relative index of Ct values normalized to GAPDH by real-time PCR. PCR was carried out using the comparative Ct method (Applied Biosystems software) with SYBR Green PCR core reagents (Applied Biosystems) and analyzed using Applied Biosystems 7900HT Real-Time PCR System software SDS 2.2.1. All primers used were specific to mouse. *denotes p value < 0.01 compared to WT + OVA values. # denotes p value < 0.01 compared to WT + alum (control baseline values). $n = 5$/group pooled from two experiments. Expression of the gene of interest was expressed in relative values normalized to the values obtained for mouse GAPDH. Compared to post-OVA WT values, Rho kinase RGS-5 mRNA was decreased 2.3-folds and MMp-10 threefold, while MMP-9 was increased eightfold and MMP-28 increased 2.3-fold (*denotes p value < 0.05 compared to post-OVA WT)

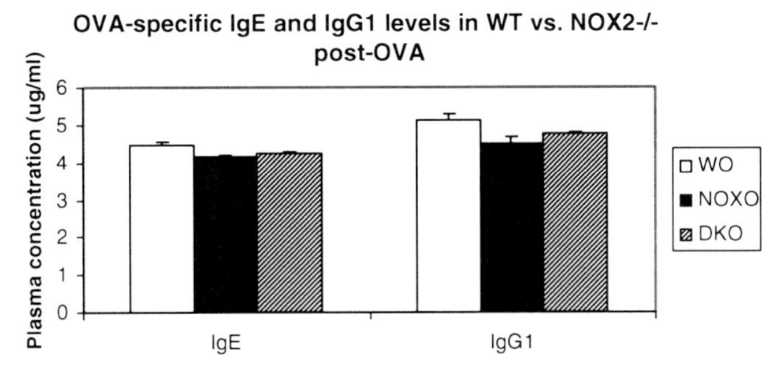

Fig. 3.32 *Plasma concentrations of OVA-specific IgE and IgG1 are comparable between Wt and KO mice post-OVA*. OVA-specific IgE and IgG1 were measured by standard ELISA using a capture and a detection antibody for each of the immunoglobulins. Data showed represents mean of values pooled from three independent experiments ($n = 5$/group). Abbreviations used are as follows: *WT* wild type, *NOX* gp91phox−/−, *DKO* gp91phox-MMP12 double knockout, *WO* WT+OVA, *NOXO* gp91phox−/−+OVA, and *DKOO* gp91phox-MMP12 double knockout+OVA. Levels of OVA-specific cytokines in the alum-treated corresponding control groups were undetectable

mice. With PMA/ionomycin (10 ng/ml), CD4+ (post-OVA WT) was 2.3-fold more than that in NOX−/−, while CD8+ was 1.5-fold more in post-OVA WT than in either KO mice. Overall, whereas proliferation of both T cell subsets to anti-CD3/CD28 is comparable, response to PMA/ionomycin is somewhat compromised in KO post-OVA (Fig. 3.33a, b).

Functionality of Macrophages

Macrophages and neutrophils are the ultimate downstream cells contributing to the asthma phenotype. Their functions are measured by oxidative burst response to PMA and chemotaxis to MCP-1. Figure 3.34 shows drastic downregulation of DHR + cells by FACS gated on both Gr-1+ and F4/80+ populations showing either myeloid population to be incapable of showing respiratory burst response by generating reactive oxygen species by responding to PMA. Figure 3.35, however, surprisingly shows upregulation in both gp91phox−/− and DKO alveolar macrophages post-OVA. So here is a clear forking of signaling pathways regulating proliferation on the one hand, which was decreased in the KO mice (to PMA/Ionomycin), and oxidative burst response on the other which was drastically compromised (again to PMA), showing perhaps that the protein kinase C pathway may be dysfunct in the absence of gp91phox subunit of NADPH oxidase, but when it comes to chemotaxis, there may be an upregulation of MCP-1 explaining the increase in actively migrating cells.

T-Cell-Macrophage Cross Talk

Based on the aforementioned responses of T cells and macrophages, it seems apparent that both cells are able to function well in response to OVA on their own at least as far as the asthma phenotype is concerned. They migrate in increased numbers from the blood, and resident and recruited cells are found in impressive inflammatory exudates around the airways. So the next question was whether there is efficient cross talk between the T cells upstream and the macrophages downstream. To this end we did a mixed lymphocyte reaction using first the CD4+ T cells as the responders and the γ-irradiated alveolar macrophages as the stimulators from the experimental mice themselves and then used CD4+ T cells from splenocytes of BALB/c mice. Increase in proliferation measured by MTT assay (OD 570 nm) shows increased T cell: APC interaction both when autologous APCs (macrophages from adherent cell population in BALf of the same animal) were used and then APCs from experimental animals were used as stimulators to CD4+ T cells from spleen of BALB/c mice which were the responders (Fig. 3.36).

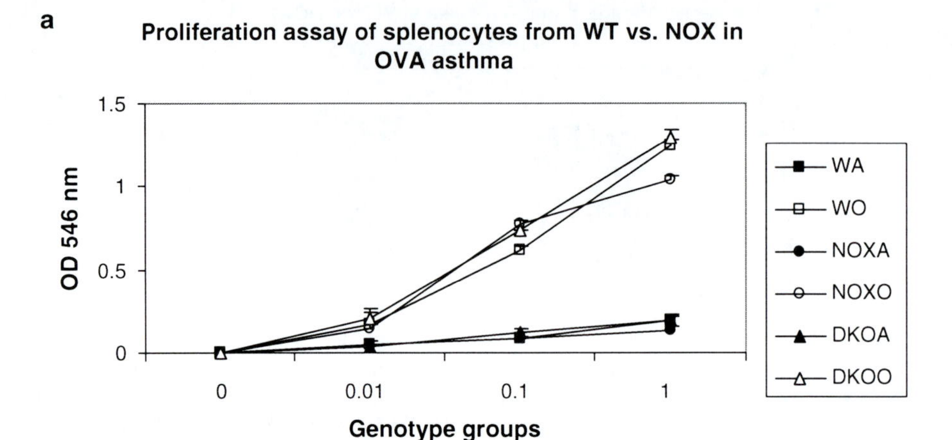

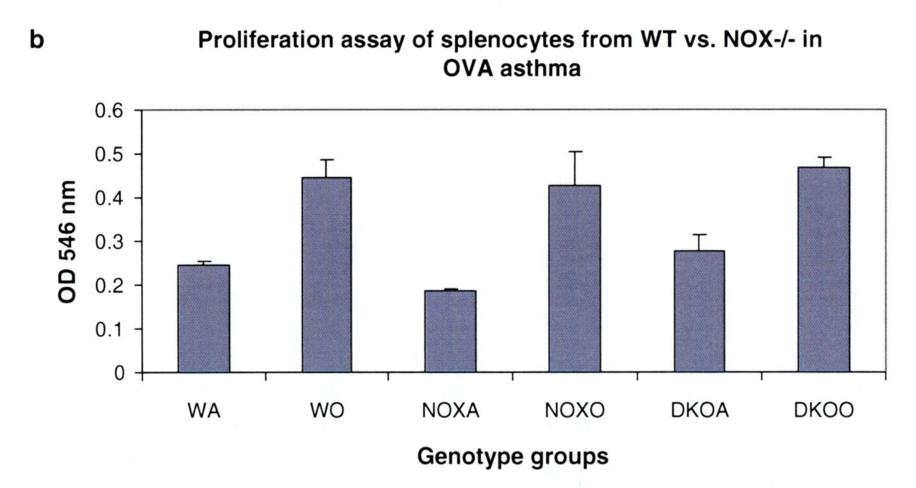

Fig. 3.33 *T cells respond similarly to stimuli in wild-type (WT) and knockout (KO) mice*. (**a**). Splenocytes from control (saline-treated) and OVA-treated mice were made into single-cell suspensions in DMEM + 10 % heat-inactivated FCS. 0.1 million cells were plated per well without and with increasing concentrations of anti-CD3 antibody and a constant concentration of anti-CD28 antibody (1 μg/ml) and cultured for 3 days. (**b**). 1μM PMA and 10 ng/ml ionomycin were used to stimulate splenocytes from the above experimental mice and proliferation measured after 3 days. To measure proliferation, MTT assay called CellTiter96 (Promega) was used. OD 546 nm is directly proportional to the number of cells in culture. Abbreviations used are as follows: *WT* wild type, *NOX* gp91phox−/−, *DKO* gp91phox-MMP12 double knockout, *WA* WT+alum, *WO* WT+OVA, *NOXA* gp91phox−/−+alum, *NOXO* gp91phox−/−+OVA, *DKOA* gp91phox-MMP12 double knockout+alum, and *DKOO* gp91phox-MMP12 double knockout+OVA. Data presented are average of three independent experiments ± SEM. ($n = 5$/group)

iNOS Expression

iNOS is a surface enzyme expressed by macrophages that are of the M1 or killer phenotype. Figure 3.37 shows decrease in percent iNOS + cells in PB, spleen, the lung, and BALf, but not BM. This may indicate that there is a skewing of macrophage phenotype from killer to healer phenotype. This corroborates well with the data in Figs. 3.34 and 3.35 in which these macrophages although migrating to the inflammatory focus in increased numbers are incapable of typical phagocytic functions, which indicates a clear dichotomy in their signaling pathways.

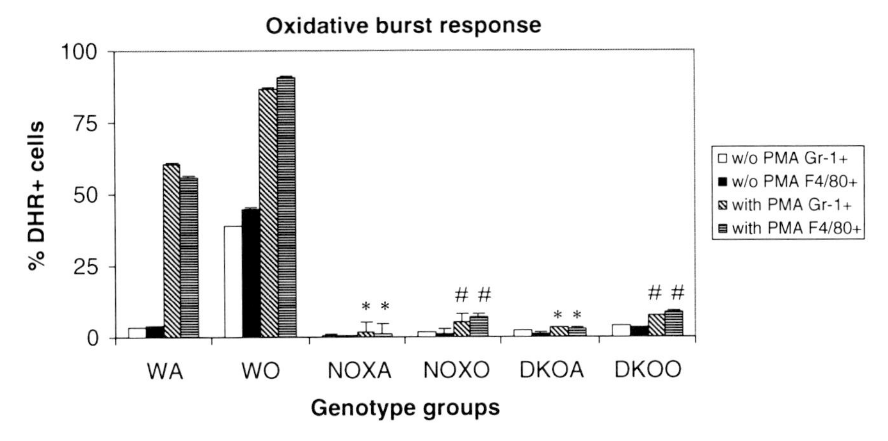

Fig. 3.34 *Myeloid cells in BALf of both KO mice fail to respond to PMA*. Alveolar leukocytes (0.5 × 106 cells) were stained with F4/80-Cy-Chrome and Gr1-APC for 30 min on ice, washed in PBS, warmed up at 370C for 5 min, and loaded with 5 mM dihydrorhodamine 123 (Molecular Probes, Eugene, OR). After 10 min at 370C, cells were split in two equal aliquots, and PMA (Sigma, St. Louis, MO) was added to one aliquot at final concentration of 1 mM. After 10 min incubation cells were washed in ice-cold PBS and immediately subjected to FACS analysis. Cells were gated on neutrophils (Gr1hi), or monocyte/macrophages (F4/80+), and percentage of cells positive for dihydrorhodamine 123 fluorescence with or without PMA treatment was determined for each gate. Results shown are mean of three independent experiments ± SEM. ($n = 5$/group). *denotes p value < 0.05 compared to WT without PMA treatment and # denotes p value < 0.05 compared to WT post-PMA treatment. While WT cells respond to PMA before as well as after OVA challenge, cells from both KO mice before as well as after OVA failed to respond appreciably. DHR was measured at fluorescent channel 1 in using a BD FACSCalibur and DHR + cells (CD45 + gated and Gr-1+ gated or F4/80+ gated) were analyzed using CellQuestPro software

Expression of Costimulatory Molecules

We hypothesized that expression of MHC molecule, which controls T-cell activation by APC, may be somehow affected in this mechanism. Figure 3.38 indicates a 1.65-fold increase in post-OVA gp91phox−/− lung parenchyma cells and a 1.38-fold increase in DKO cells in B7.1 positive cells from undetected positive cells in saline treated in any group. B7.2 and MHCII expressions were however decreased in both KO mice with 3.28-fold and 3.18-fold decrease respectively in gp91phox−/− and DKO. MHCII expression was downregulated by 1.18- and 1.13-fold in the two KO mice respectively.

Bleomycin-Induced Fibrosis

Figure 3.39 shows increase in soluble collagen content of whole lung in gp91phox−/− but not DKO where values were similar to baseline showing that response did not occur. Gp91phox−/− lung showed 1.3-fold increase in freshly synthesized collagen post-OVA compared to post-OVA WT. Histopathological staining with trichrome stain showed appreciable collagen deposits in WT and gp91phox post-OVA but less in DKO lung (**Panel A**). A similar trend was found in Picrosirius red stain (**Panel C**). Interestingly TGFβ staining was pronounced around airways in WT and gp91phox−/− post-OVA but not DKO (Fig. 3.40).

Cells in BALf and Lung Post-Bleomycin over Time

Total number of cells and cell subsets was measured periodically over time in WT and the KO mice (Tables 3.14, 3.15, 3.16, and 3.17). Total cells were increased in bleomycin-treated NOX−/− mice at day 7 after bleo with macrophages and lymphocytes contributing the most of the increase. PMNs also increased in percentage but total number was comparable to that in post-bleo WT. DKO on the other hand did not show very low increase in total number of cells over their saline-treated control group with lymphocytes and PMNs contributing to the slight increase entirely. Compared to saline-treated WT,

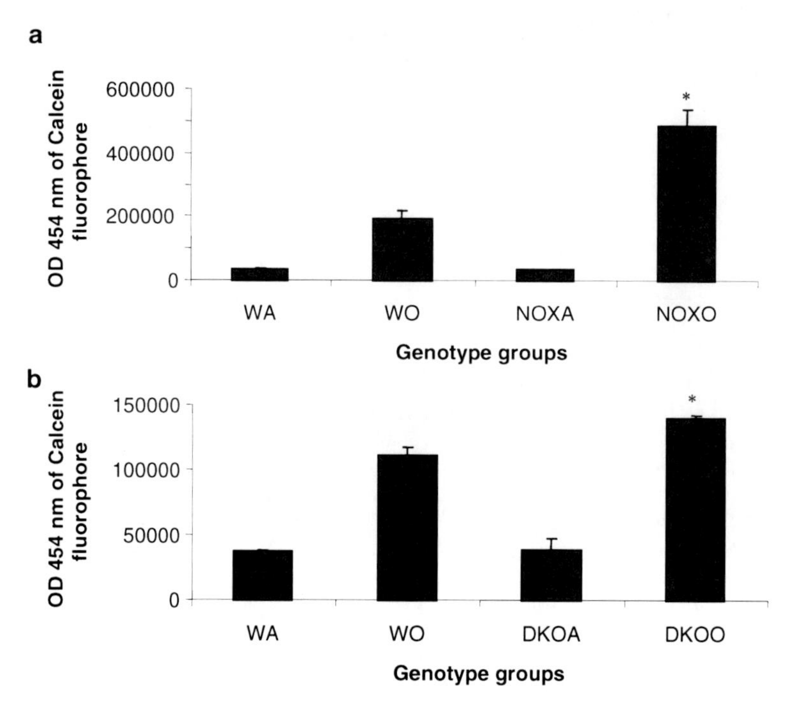

Fig. 3.35 *MCP-1-driven chemotaxis of alveolar macrophages was increased in both KO mice post-OVA*. 15 mM MCP-1 was put in 29 μl volume in the lower well and 10 × 106 alveolar macrophages (from 4 mice/experimental group) and also in 29 μl volume in the upper wells of a 96-well Neuroprobe CTX plates (Chemicon) in high glucose medium for 2 h followed by detachment by mechanical scraping and resuspension in phenol red-free high glucose DMEM (Gibco) with 5 % FBS with 0.5 μg/ml Calcein-AM (1:2,000 dilution) and incubation for 20 min at 370C. Migrated cells were quantified by fluorescence (excitation at 488 nm, emission at 520 nm) using a Victor 3 V (Perkin Elmer laboratories) using a Wallac1420 software. 2.5-fold and 1.26-fold increase in OD (proportionate to the number of fluorescing cells in the upper well equivalent to the number of cells migrated) was found in post-OVA gp91phox−/− and DKO mice respectively. *denotes p value < 0.05 compared to values in OVA-treated wild-type group

number of cells increased 6.6-folds a week after bleomycin treatment. This denuded at day 14 and further at day 21. So similar to the previous data, cell number in BALf is actually not an indicator of the extent of fibrosis by day 21. Similar trends were found in the lung (Table 3.16). D7 probably signals onset of fibrosis by increased cellular recruitment. Macrophages and T cells are the chief secretors of TGFβ traditionally thought to activate collagen synthesis and deposition by alveolar epithelial cells. Days 8–21 therefore is the scar tissue formation period when AEI have become denuded and so have AEII. This is shown in Table 3.17 where there is a progressive decrease in AEI through day 21 while AEII first increased slightly only to equilibrate at day 21. NOX−/− BALf showed a 1.7-fold increase in cell number which decreases predictably by day 21. DKO BALf however shows no appreciable increase over saline treated either at day 7 or at day 21. Macrophages seem to be the chief cell populations accounting for this increase. In the lung (Table 3.16), however, on day 7, a similar trend to that in BALf is found. The only difference is that both macrophages and lymphocytes make up for the increase.

Discussion

Briefly, deletion of gp91^{phox} results in enhancement of composite asthma phenotype in mouse. Double deletion of gp91phoix and MMp12, a critical enzyme for phagocyte-associated

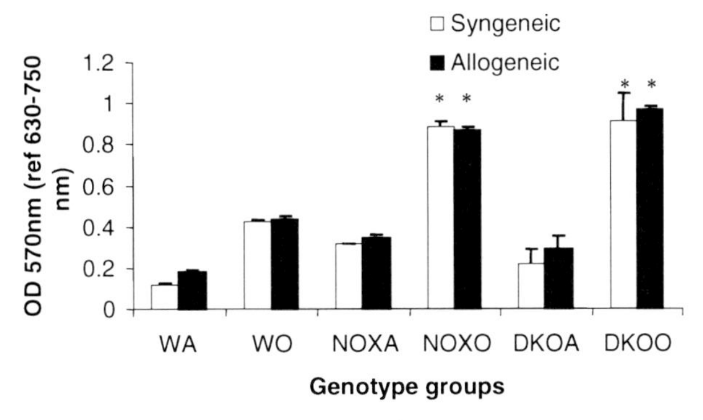

Fig. 3.36 *Mixed lymphocyte reaction with CD4+ T cells and antigen-presenting cells (APC) before and after OVA treatment.* 0.1 × 105 CD4+ T cells isolated by magnetic activated cell separation (MACS) by positive selection from spleen of the mice were co-cultured with γ-irradiated 0.1 × 103 adhering alveolar macrophages from BALf of the same experimental animal. Both cell types were from the experimental animals themselves, viz., C57Bl/6 WT and KO mice. This is syngeneic MLR where the APCs (alveolar macrophages) were γ-irradiated (3300rads). Allogeneic MLR reaction involves co-culturing 0.1 × 105 CD4+ T cells from the spleen of BALB/c mice with 0.1 × 103 γ-irradiated APCs from BALf of the experimental mice. Both control (saline-treated) and OVA-treated of each group were tested. Each culture was incubated with 1 μg/ml OVA. The responders here are the T cells and the stimulators are the APCs, viz., alveolar macrophages which are γ-irradiated to inhibit their own proliferation. Since acute allergic asthma is a Th2-mediated phenomenon, interaction between T cells and macrophages will elucidate functional cross talk between the two cell types when responders are autologous as well as when they are from a different species. Twofold increase in post-OVA NOX versus post-OVA WT and 2.16-fold increase in post OVA-DKO versus post-OVA show that T cell:APC interaction is actually more efficient in the absence of the gp91phox and MMP12 as well as gp91phox

inflammation results in no alteration of the phenotype generated in the single deletion of gp91phox−/−. There is a clear dichotomy in the macrophage function, but overall, T-cell function and macrophage functions are well coordinated. While inflammation in terms of inflammatory cell migration is enhanced, macrophage function downstream may not be all that bad. If gp91phox is imagined to have a regulatory role in the development of the asthma phenotype, MMP12 seems to have a similar if not synergistic effect. These molecules are however not necessary for migration, but are critical for PMA-induced proliferation and MCP-1-induced chemotaxis. The overall Th2 response was enhanced possibly due to a lack of control over T cell: APC cross talk in the KO mice as shown by MLR. Increased B7.1 but decreased B7.2 and MHCII expression may provide possible mechanisms for this regulatory function of gp91phox and MMP12. On the other hand, in a predominantly macrophage, perpetrated disease model of bleomycin-induced pulmonary fibrosis, gp91phox−/− mouse shows the same trend of runaway inflammation and the resultant fibrosis but not so when MMP12 is deleted additionally. This may mean that while deletion of gp91phox may abolish the control mechanism required for keeping the effects of bleomycin under check, additional deletion of MMP12 fails to elicit any effect of bleomycin after 21 days of treatment when fibrosis is fully developed. Therefore, MMP12 may be critical for phagocytes to process signals necessary for (a) increased AEI and AEII mortality, (b) increased collagen synthesis, and (c) inflammation, the first step for accumulation of the cell populations in the lung and BALf to initiate the onset and development of the fibrotic process itself.

The results described above indicate that gp91phox−/− mice respond to OVA in a more exaggerated fashion compared to WT post-OVA, in terms of the total number of cells migrated to the lung (1.8-folds) and BALf (1.7-folds), although total number of cells in bone marrow (obtained from two femurs) and that in circulating peripheral blood was comparable. Expressed as a fraction of circulating cells

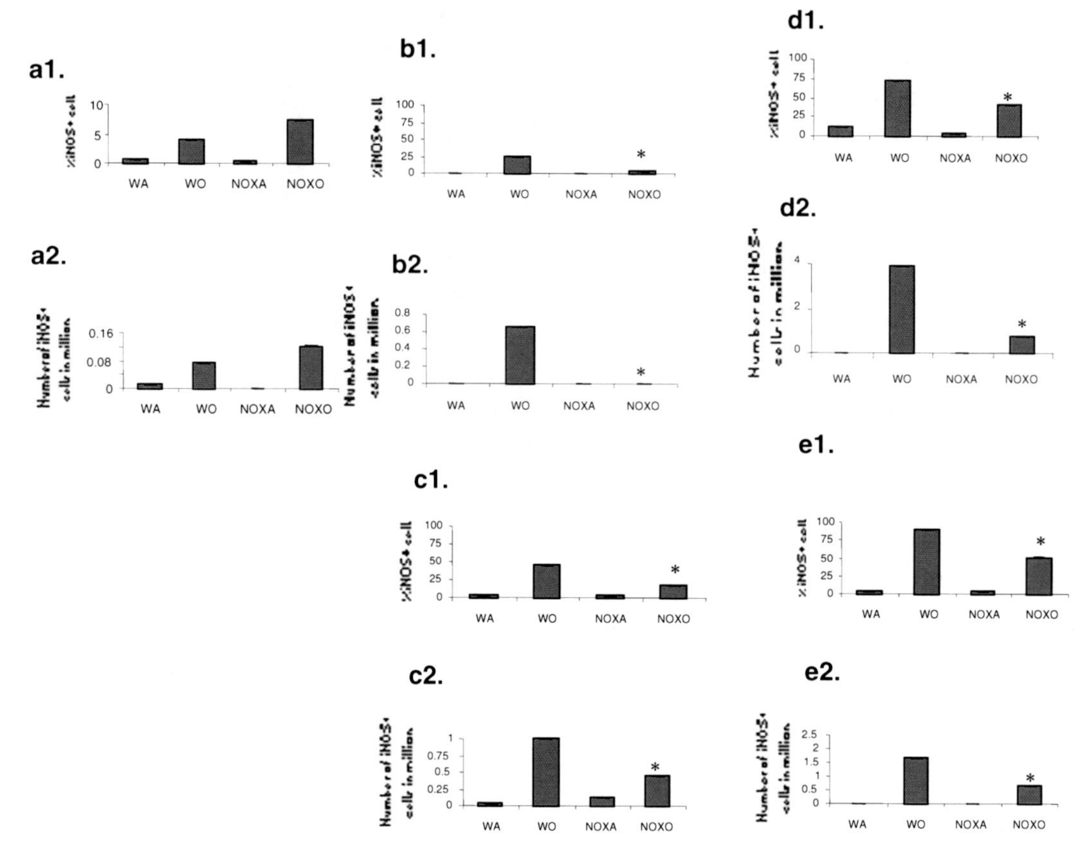

Fig. 3.37 *Decrease in percent iNOS positive cells as well as total number of them in gp91phox−/− (NOX) mice post-OVA compared to wild-type (WT) post-OVA*. Cells from all tissues, viz., BM, PB, spleen, lung, and BALf of NOX−/− versus WT with and without OVA. Or we could put % and # in tabular form. **A1–E1**. Percent iNOS positive cells in BM, PB, spleen, LP, and BALf, respectively. **A2–E2**. Number of iNOS positive cells in BM, PB, spleen, LP, and BALf respectively cells in million. Data shows the mean of two independent experiments which were pooled ± SEM. (*denotes p value < 0.05 compared to WO). Abbreviations used are as follows: *WT* wild type, *NOX* gp91phox−/−, *WA* WT+alum, *WO* WT+OVA, *NOXA* gp91phox−/−+alum, *NOXO* gp91phox−/−+OVA

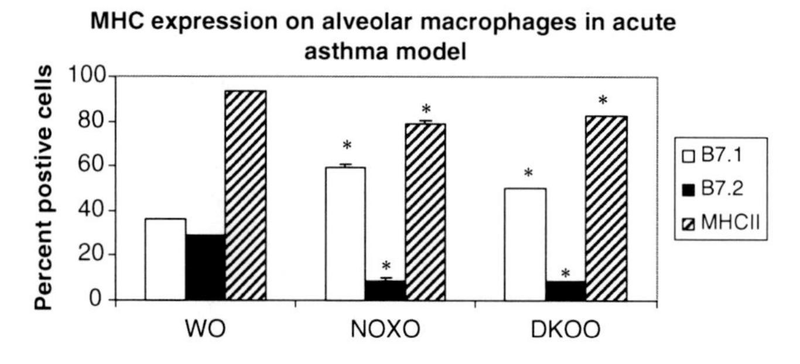

Fig. 3.38 *Increase in B7.1 but decrease in B7.2 and MHCII expression in BAL cells*. B7.1 and B7.2 are costimulatory molecules expressed on alveolar macrophages and other antigen presentation cells like the dendritic cells and also B cells and monocytes. Expression of the said molecules was measured by FACS using specific fluorochrome-conjugated antibodies from Pharmingen. The data presented shows percent cells positive for the given antigen, expressed as mean ± SEM. $n = 4$/group

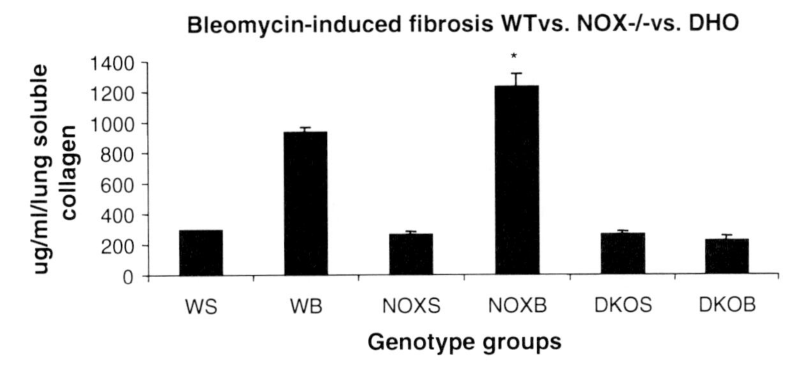

Fig. 3.39 *Increase in collagen content of lung post-bleomycin in gp91phox−/− but not in MM12-gp91phox double knockout mice.* Total collagen content freshly synthesized in the lung was measured by Sircol assay. Data represents the mean of two independent experiments, $n = 4$/group. Data are expressed as average ± SEM. *indicates p value < 0.05 compared to post-OVA value in wild-type mice. Abbreviations are as follows: *WS* saline-treated wild type (baseline), *WB* bleomycin-treated (single intra-tracheal dose of 0.075 U/ml bleomycin and animals were sacrificed 14 and 21 days later. Here day 21 data are shown), *NOXS* saline-treated gp91phox null, *NOXB* bleomycin-treated gp91phox null, *DKOS* saline-treated MMP12- gp91phox double knockout, and *DKOB* bleomycin-treated MMP12- gp91phox double knockout

recruited into the lungs and BALf in response to allergic immune response, both knockout mice seem to show similar trends. Cell subsets, for which recruitment index is more than 1, indicate a cumulative effect where both cells from circulation, as well as resident cells normally present in the pulmonary milieu in surveillance, seem to be equally important. Recruitment of B cells, monocytes, neutrophils, and basophils is increased in the lungs of both knockout mice compared to post-OVA wild type, while that of T cells, neutrophils, and basophils in BALf is increased in the knockout versus the OVA-treated wild type (Table 3.10).

MMP-9 is known to play a role in the migration and inflammatory responses by Eos and PMN, and control repair responses in asthma, especially in the resolution of allergic inflammation. So upregulation of MMP-9 gene in upon gp91phox deletion and also post-OVA treatment of the knockout mouse cannot be explained easily. It may either have a protective role which is confounded upon deletion of gp91phox or it may be a compensatory mechanism in reaction to the deletion of the same. Any complementary role of this metalloprotease with the subunit of NADPH oxidase is unknown. MMP-12, on the other hand, controls migration of monocytes and macrophages to inflammatory sites and airway remodeling by degrading ECM proteins [134]. It is supposed to have a protective effect in emphysema [135].

B7.1, a costimulatory signal necessary for the activation of T cells, can be expressed on cell surface by B cells, dendritic cells, and macrophages, the antigen-presenting cells. It is associated with activation of cell-mediated response especially Th2 response. At baseline, they are not expressed but upon activation are upregulated. In our model, upregulation of B7.1 but downregulation of B7.2 and MHCII shows a possible mechanism by which gp91phox and MMP12 may synergistically regulate Th2 responsiveness, and deletion of the same disrupts this pathway.

Mature T lymphocytes become activated to perform their effector functions when stimulated by appropriate APC bearing MHC class I or class II molecules. They require both antigen-specific and immunoregulatory signals for optimum activation. The first signal comes from the T-cell receptor as it recognizes antigenic peptides presented by MHC molecules. A variety of costimulatory receptors provide the second signal when they interact with ligands on APCs. The B7 family of ligands and their receptors

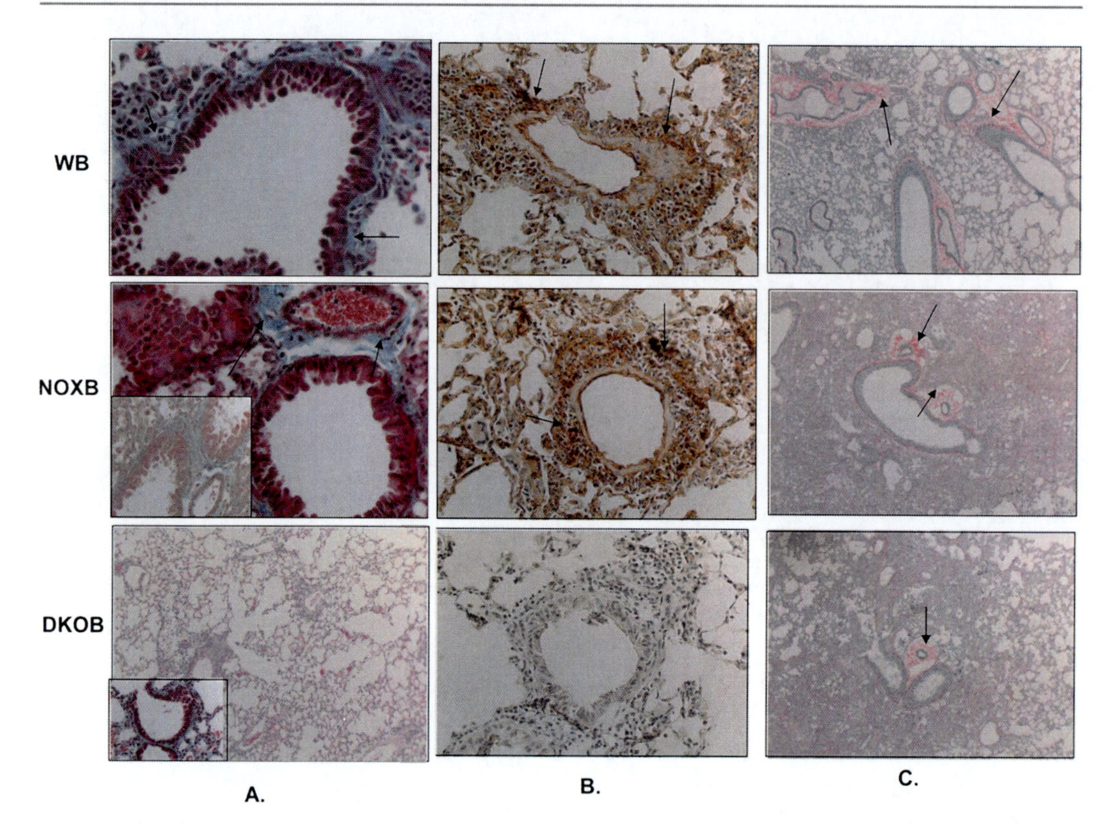

Fig. 3.40 *Increased inflammation and fibrosis in the lung of WT and gp91phox−/− mice but not MMP12gp91phox double knockout mice.* Twenty-one days post-bleomycin treatment, mice were sacrificed and lungs were fixed overnight in 4 % paraformaldehyde and embedded in paraffin. 8-μ sections were stained by dyes (A, C. Masson's trichrome stain and Picrosirius red, respectively) and immunostained with TGFβ antibody and color developed by DAB (B). Arrows point at blue collagen deposits in panel A, TGFβ positive stain in panel B, and to red collagen staining in panel C. In panel A, pinkish purple stains are that of cytoplasm and blue are the nuclei of inflammatory cells. Light blue stains are of collagenous deposits. While WT and gp91phox−/− (designated as NOX−/−) show increased collagen deposits and inflammation around airways post-OVA (all sections are those of post-OVA lung), MMP-12-gp91phox double knockout lung shows minimum collagen deposition as well as TGFβ expression. While Panel A and B were at 60× magnification, Panel C was at 10×. Panel A inset shows another area of collagen deposition in bleo-treated wild-type (WB), NOXB and DKOB are bleo-treated gp01 phox−/− and double knockout respectively. The photomicrographs were taken in an Olympus VX4L microscope and Nikon digital camera. All photomicrographs were at 60× magnification

is important in pathways of costimulation and inhibition of immune responses. B7.1 and B7.2 are costimulatory molecules capable of turning the immune system on by regulating antigen-specific activation and proliferation of lymphocytes and play a critical role in regulating the immune system. Appropriate costimulation may alter the number and strength of MHC-peptide-TCr complexes necessary for T-cell activation. It may also complement the intracellular signals sent via T-cell receptors and help T cells adhere to APCs to promote MHC-peptide-TCR interaction. These MHC molecules bind TCR and simultaneously engage either CD8 or CD4. T-cell activation and differentiation requires not only TCR recognition of the antigen-MHC complex but also costimulation through the interaction of accessory molecules on APC and their corresponding receptors on T cells. CD28 is an important costimulatory receptor on T cells and binds to CD80 (B7-1) and CD86 (B7-2) on activated APC [136, 137]. The crucial role played by CD28 in

Table 3.14 Trends in cellular migration to BALf and lung post-bleomycin over time and number of constitutive cells in the lung. Number of cells in BALf after a week of bleomycin treatment

	WT saline (day 7)	WT bleo (day 7)	NOX saline (day 7)	NOX bleo (day 7)	DKO saline (day 7)	DKO bleo (day 7)
Total cell content ($\times 10^5$/ml)	4.27 ± 0.184	28.3 ± 0.224	4.515 ± 0.223	48.91* ± 1.59	2.61 ± 0.16	8.42 ± 0.13
Macrophages (×/ml)	4.16 ± 0.57(97.2 %)	11.14 ± 4.08(37.6 %)	4.01 ± 0.31(89.9 %)	28.14* ± 6.83(57.47 %)	1.87 ± 0.64(71.64 %)	4.19 ± 3.18(49.82 %)
Lymphocytes (×/ml)	0.08 ± 0.01(2.0 %)	12.63 ± 2.49(46.0 %)	0.09 ± 0.02(2.01 %)	16.91* ± 3.65(34.53 %)	12 ± 0.01(4.59 %)	1.44 ± 0.66(17.12 %)
Neutrophils ($\times 10^5$/ml)	0.03 ± 0.01(0.7 %)	4.5 ± 0.97(16.3 %)	0.11 ± 0.01(0.02 %)	3.91 ± 0.53(7.98 %)	0.63 ± 0.14(24.13 %)	2.78 ± 1.06(33 %)

Data represent CD45+ hematopoietic cells
Number of cells in BALf a week after bleomycin treatment
Data presented are mean ± SEM of 2 independent experiments ($n = 5$/group). CD45+ cells were gated and macrophages (Gr1-F4/80hi), lymphocytes (CD3+ and β220+), and neutrophils (Gr-1hi F4/80-) were identified, their percentages analyzed by CellQuestPro and their total numbers calculated from the total number of cells obtained by lavage as described in materials and methods
Data presented is pooled from 2 independent experiments ± SEM
Abbreviations used are as follows: *WS* wild type (WT) + saline (day 7), *WB7* WT + post-bleomycin day 7, *WB14* WT + post-bleomycin day 14, *Wβ21* WT + post-bleomycin day 21, *NOXS* gp91phox−/− + saline day 21, *NOXβ21* gp91phox−/− + bleomycin day 21, *DKOS* gp91phox-MMP12 double knockout + saline (day 21), *DKOβ21* gp91phox-MMP12 double knockout + bleomycin (day 21)
*denotes p value < 0.05 compared to WT post-bleo on corresponding time points

the proliferation and differentiation of T cells has been highlighted by studies using CD28 knockout mice. However, some immune responses remain intact in the absence of CD28, perhaps because a prolonged TCR signal overcomes the need for costimulation and/or other costimulatory molecules can substitute for CD28 [138, 139]. Additional T-cell costimulatory molecules, i.e., ICOS, were identified and have functions similar to, but not completely overlapping with, CD28. ICOS is not expressed by resting T cells, but is induced after T-cell activation and is important for T-cell activation and effector function [139, 140]. ICOS costimulation failed to significantly upregulate IL-2; however, it upregulated the secretion of IL-4, IL-5, IFN-γ, TNF-α, and GM-CSF to 50–70 % of the levels achieved with CD28 costimulation [141]. Following cellular activation, lymphocytes can differentiate into Th1 or Th2 cells according to the type of cytokine they produce, the balance of which will ultimately determine the outcome of the cellular response. Typically, production of IFN-γ and IL-12 will favor a Th1 response, whereas presence of IL-4 and IL-10 will determine a Th2 pattern. In HP no clear pattern of polarization has been defined. In experimental models, Th1 responses may be important since Th1 $CD4^+$ cells can adoptively transfer the disease to healthy animals [142]; IFN-γ and IL-12 may also play a role in the pathogenesis of HP in mice [143, 144], but in human HP, recent studies suggest that a Th2-type response is predominant [145]. T-cell activation requires at least two distinct signals [146]; the first is Ag specific and is delivered through the engagement of TCRs. The second signal is mediated by the interaction of costimulatory molecules present on APCs with their ligands on T cells. The B7:CD28/CTLA4 is a major pathway which provides these potent signals, crucial for complete T-cell activation. CD28 and CTLA4 are ligands for B7-1 and B7-2. These ligands bind to both B7 but with different avidities; CTLA4 binding is 20– 100-fold higher than CD28. This difference in avidity has been exploited to block B7-CD28 interactions by the use of CTLA4-Ig, a soluble fusion protein made from the extracellular portion of CTLA4 linked to the Fc portion of IgG [147]. Furthermore, B7/CD28 costimulatory pathway may influence not only

Table 3.15 Trends in cellular migration to BALf and lung post-bleomycin over time and number of constitutive cells in the lung. Number of CD45 + cells in BALf before and after bleomycin treatment (1 i.t. dose of 0.074U/ml in 40μl volume)

BALf	Mean							
10E5/ml	WS	WB7	WB14	WB21	NOXS	NOXB21	DKOS	DKOB21
Total cells	4.27 ± 0.18	28.3 ± 0.22	10.32 ± 2.74	9.32 ± 1.74	4.51 ± 0.23	8.36 ± 1.32	4.55 ± 0.16	5.03 ± 1.94
Macs	4.16 ± 0.57	11.14 ± 4.08	8.91 ± 4.93	7.06 ± 3.92	3.86 ± 1.07	6.91 ± 1.96	3.91 ± 0.74	4.03 ± 0.43
Lympho	0.08 ± 0.01	12.63 ± 2.49	1.76 ± 0.67	1.33 ± 0.67	1.32 ± 0.54	2.43 ± 0.43	1.44 ± 0.32	0.56 ± 0.21
PMN	0.03 ± 0.01	4.5 ± 0.97	0.41 ± 0.07	0.87 ± 0.32	0.67 ± 0.32	1.02 ± 0.22	0.67 ± 0.11	0.41 ± 0.13

Data represent CD45+ hematopoietic cells

Number of cells in BALf before and after bleo challenge where WT mice were assessed at days 7, 14, and 21 and KO mice were assessed on days 7 and 21

Data presented are mean ± SEM of 2 independent experiments ($n = 5$/group). CD45+ cells were gated and macrophages (Gr1-F4/80hi), lymphocytes (CD3+ and β220+), and neutrophils (Gr-1hi F4/80-) were identified, their percentages analyzed by CellQuestPro and their total numbers calculated from the total number of cells obtained by lavage as described in materials and methods

Data presented is pooled from 2 independent experiments ± SEM

Abbreviations used are as follows: *WS* wild type (WT) + saline (day 7), *WB7* WT + post-bleomycin day 7, *WB14* WT + post-bleomycin day 14, *Wβ21* WT + post-bleomycin day 21, *NOXS* gp91phox−/− + saline day 21, *NOXβ21* gp91phox−/− + bleomycin day 21, *DKOS* gp91phox-MMP12 double knockout + saline (day 21), *DKOβ21* gp91phox-MMP12 double knockout + bleomycin (day 21)

*denotes p value < 0.05 compared to WT post-bleo on corresponding time points

Table 3.16 Trends in cellular migration to BALf and lung post-bleomycin over time and number of constitutive cells in the lung. Number of (CD45+) cells in LP before and after bleomycin treatment

LP (da 7)	Mean					
10E5/ml	WS	WB	NOXS	NOXB	DKOS	DKOB
Total cells	7.27 ± 0.07	38.3 ± 4.04	8.86 ± 0.08	54.19* ± 2.6	6.85 ± 0.27	9.56 ± 0.65
Macs	6.16 ± 0.57	22.14 ± 4.08	7.13 ± 2.43	31.38* ± 11.32	5.41 ± 1.43	6.36 ± 2.71
Lympho	1.08 ± 0.01	9.63 ± 2.49	0.97 ± 0.12	21.41* ± 4.93	0.93 ± 0.21	2.47 ± 0.67
PMN	0.03 ± 0.01	7.5 ± 0.97	0.72 ± 0.04	1.34 ± 0.15	0.51 ± 0.03	0.73 ± 0.33

Data represent CD45+ hematopoietic cells

Number of cells in lung parenchyma (Lp) on day 7 in all as no appreciable difference in cell numbers were found over time from day 7 to day 21

Data presented are mean ± SEM of 2 independent experiments ($n = 5$/group). CD45+ cells were gated and macrophages (Gr1-F4/80hi), lymphocytes (CD3+ and β220+), and neutrophils (Gr-1hi F4/80-) were identified, their percentages analyzed by CellQuestPro and their total numbers calculated from the total number of cells obtained by lavage as described in materials and methods

Data presented is pooled from 2 independent experiments ± SEM

Abbreviations used are as follows: *WS* wild type (WT) + saline (day 7), *WB7* WT + post-bleomycin day 7, *WB14* WT + post-bleomycin day 14, *Wβ21* WT + post-bleomycin day 21, *NOXS* gp91phox−/− + saline day 21, *NOXβ21* gp91phox−/− + bleomycin day 21, *DKOS* gp91phox-MMP12 double knockout + saline (day 21), *DKOβ21* gp91phox-MMP12 double knockout + bleomycin (day 21)

*denotes p value < 0.05 compared to WT post-bleo on corresponding time points

the extent of T-cell activation but also the regulation of T-cell differentiation [148]. Depending on the system studied, B7 costimulation has been shown to influence both Th1 and Th2 cytokine production [149]. In the normal lung, alveolar macrophages (AM) have a low expression of B7 molecules and a poor capacity to function as APCs [150]. So antagonistic alterations in B7 family of receptors in the acute asthma pathway may indicate a definite role for either gp91phox or both gp91phox and MMP12 in controlling the costimulatory activating pathway in T-cell activation in Th2 response. Role of MMP12 and gp91phox is however probably completely different in bleomycin-induced fibrosis and needs further study.

Table 3.17 Trends in cellular migration to BALf and lung post-bleomycin over time and number of constitutive cells in the lung. Number of CD45- cells counterstained with alveolar epithelial cells before and after bleomycin treatment

	CD45−			
	Saline	Bleomycin		
	Day 21	Day 7	Day 14	Day 21
WT				
AEI	95.7 ± 4.87	79.47 ± 2.86	67.43 ± 1.76	58.41 ± 4.83
AEII	4.3 ± 1.76	9.41 ± 1.86	8.66 ± 1.07	5.96 ± 1.12
NOX−/−				
AEI	95.7 ± 3.87	ND	ND	58.41 ± 3.97
AEII	4.3 ± 1.12	ND	ND	5.96 ± 0.56
DKO				
AEI	95.7 ± 4.16	ND	ND	97.41 ± 3.87
AEII	4.3 ± 0.56	ND	ND	4.96 ± 1.87

Data represent CD45- cells staining positive for alveolar epithelial cells
Number of cells in BALf of saline-treated and post-bleomycin (day 7) wild type (WT), gp91phox−/− (NOX) and MMP12- gp91phox double knockout (DKO) mice were assessed by Z1 Coulter particle counter from Beckman Coulter and differential counts done by specific fluorochrome antibody conjugates by FACS and by morphology in light microscope
Data presented are mean ± SEM of 2 independent experiments ($n = 5$/group). CD45+ cells were gated and macrophages (Gr1-F4/80hi), lymphocytes (CD3+ and β220+), and neutrophils (Gr-1hi F4/80-) were identified, their percentages analyzed by CellQuestPro and their total numbers calculated from the total number of cells obtained by lavage as described in materials and methods
Data presented is pooled from 2 independent experiments ± SEM
Abbreviations used are as follows: *WS* wild type (WT) + saline (day 7), *WB7* WT + post-bleomycin day 7, *WB14* WT + post-bleomycin day 14, *Wβ21* WT + post-bleomycin day 21, *NOXS* gp91phox−/− + saline day 21, *NOXβ21* gp91phox−/− + bleomycin day 21, *DKOS* gp91phox-MMP12 double knockout + saline (day 21), *DKOβ21* gp91phox-MMP12 double knockout + bleomycin (day 21)
*denotes p value < 0.05 compared to WT post-bleo on corresponding time points

References

1. Slattery D, Wong SW, Colin AA. Levalbuterol hydrochloride. Pediatr Pulmonol. 2002;33:151–7.
2. Cho SH, Hartleroad JY, Oh CK. (S)-Albuterol increases the production of histamine and IL-4 in mast cells. Int Arch Allergy Immunol. 2001;124:478–84.
3. Baramki D, Koester J, Anderson AJ, Borish L. Modulation of T-cell function by (R)- and (S)-isomers of albuterol: anti-inflammatory influences of (R)-isomers are negated in the presence of the (S)-isomer. J Allergy Clin Immunol. 2002;109:449–54.
4. Agrawal DK, Ariyarathna K, Kelbe PW. (S)-Albuterol activates proconstrictory and pro-inflammatory pathways in human bronchial smooth muscle cells. J Allergy Clin Immunol. 2004;113:503–10.
5. Keir S, Page C, Spina D. Bronchial hyperresponsiveness induced by chronic treatment with albuterol: role of sensory nerves. J Allergy Clin Immunol. 2002;110:388–94.
6. Eaton EA, Walle UK, Wilson HM, Aberg G, Walle T. Stereoselective sulphate conjugation of salbutamol by human lung and bronchial epithelial cells. Br J Clin Pharmacol. 1996;41:201–6.
7. Boulton DW, Fawcett JP. Enantioselective disposition of salbutamol in man following oral and intravenous administration. Br J Clin Pharmacol. 1996;41:35–40.
8. Henderson Jr WR, Lodewick MJ. Animal models of asthma. In: Adkinson Jr NF, Yuninger JW, Busse WW, Bochner BS, Holgate ST, Simons FER, editors. Middleton's allergy: principles and practice. 6th ed. St Louis: Mosby; 2003. p. 465–81.
9. Henderson Jr WR, Lewis DB, Albert RK, Zhang Y, Lamm WJE, Chiang GKS, et al. The importance of leukotrienes in airway inflammation in a mouse model of asthma. J Exp Med. 1996;184:1483–94.
10. Zhang Y, Lamm WJE, Albert RK, Chi EY, Henderson Jr WR, Lewis DB. Influence of the route of allergen administration and genetic background on the murine allergic pulmonary response. Am J Respir Crit Care Med. 1997;155:661–9.
11. Johnson M. Effects of β2-agonists on resident and infiltrating inflammatory cells. J Allergy Clin Immunol. 2002;110(suppl):S282–90.
12. Wang CH, Lin HC, Lin CH, Yu CT, Liu SL, Huang KH, et al. Effect of theophylline and specific phosphodiesterase IV inhibition on proliferation and apoptosis of progenitor cells in bronchial asthma. Br J Pharmacol. 2003;138:1147–55.

13. Rogers DF. Pharmacological regulation of the neuronal control of airway mucus secretion. Curr Opin Pharmacol. 2002;2:249–55.
14. Frohock JI, Wijkstrom-Frei C, Salathe M. Effects of albuterol enantiomers on ciliary beat frequency in ovine tracheal epithelial cells. J Appl Physiol. 2002;92:2396–402.
15. Bennett WD. Effect of b-adrenergic agonists on mucociliary clearance. J Allergy Clin Immunol. 2002;110(suppl):S291–7.
16. Temann U-A, Prasad B, Gallup MW, Basbaum C, Ho SB, Flavell RA, et al. A novel role for murine IL-4 in vivo: induction of MUC5AC gene expression and mucin hypersecretion. Am J Respir Cell Mol Biol. 1997;16:471–8.
17. Grunig G, Warnock M, Wakil AE, Venkayya R, Brombacher F, Rennick DM, et al. Requirement for IL-13 independently of IL-4 in experimental asthma. Science. 1998;282:2261–3.
18. Henderson Jr WR, Chi EY, Maliszewski CR. Soluble IL-4 receptor inhibits airway inflammation following allergen challenge in a mouse model of asthma. J Immunol. 2000;164:1086–95.
19. Nguyen C, Teo J-L, Matsuda A, Eguchi M, Chi E, Henderson Jr WR, et al. Chemogenomic identification of Ref-1/AP-1 as a novel therapeutic target for asthma. Proc Natl Acad Sci U S A. 2003;100: 1169–73.
20. Burke TF, Casolaro V, Georas SN. Characterization of P5, a novel NFAT/AP-1 site in the human IL-4 promoter. Biochem Biophys Res Commun. 2000;270:1016–23.
21. Macian F, Garcia-Rodriguez C, Rao A. Gene expression elicited by NFAT in the presence or absence of cooperative recruitment of Fos and Jun. EMBO J. 2000;19:4783–5.
22. Fallon PG, Jolin HE, Smith P, Emson CL, Townsend MJ, Fallon R, et al. IL-4 induces characteristic Th2 responses even in the combined absence of IL-5, IL-9, and IL-13. Immunity. 2002;17:7–17.
23. Peterson BT, Miller EJ. Effects of enantiomers of albuterol on lung epithelial permeability. Am J Respir Crit Care Med. 2000;161:A416.
24. Petak F, Habre W, Donati YR, Hantos Z, Barazzone-Argiroffo C. Hyperoxia-induced changes in mouse lung mechanics: forced oscillations vs. barometric plethysmography. J Appl Physiol. 2001;90: 2221–30.
25. Lundblad LK, Irvin CG, Adler A, Bates JH. A reevaluation of the validity of unrestrained plethysmography in mice. J Appl Physiol. 2002;93: 1198–207.
26. Mitzner W, Tankersley C. Interpreting Penh in mice. J Appl Physiol. 2003;94:828–31.
27. Hamelmann E, Schwarze J, Takeda K, Oshiba A, Larsen GL, Irvin CG, et al. Noninvasive measurement of airway responsiveness in allergic mice using barometric plethysmography. Am J Respir Crit Care Med. 1997;156:766–75.
28. Justice JP, Shibata Y, Sur S, Mustafa J, Fan M, Van Scott MR. IL-10 gene knockout attenuates allergen-induced airway hyperresponsiveness in C57BL/6 mice. Am J Physiol Lung Cell Mol Physiol. 2001;280:L363–8.
29. Dohi M, Tsukamoto S, Nagahori T, Shinagawa K, Saitoh K, Tanaka Y, et al. Noninvasive system for evaluating the allergen-specific airway response in a murine model of asthma. Lab Invest. 1999;79: 1559–71.
30. Sartori C, Fang X, McGraw DW, Koch P, Snider ME, Folkesson HG, et al. Selected contribution: long-term effects of β2-adrenergic receptor stimulation on alveolar fluid clearance in mice. J Appl Physiol. 2002;93:1875–80.
31. Henderson Jr WR, Tang L-O, Chu S-J, Tsao S-M, Chiang GKS, Jones F, et al. A role for cysteinyl leukotrienes in airway remodeling in a mouse asthma model. Am J Respir Crit Care Med. 2002;165:108–16.
32. Oh SW, Chong IP, Dong Keun L, Jones F, Chiang GKS, Kim HO, et al. Tryptase inhibition blocks airway inflammation in a mouse asthma model. J Immunol. 2002;168:1992–2000.
33. Iio J, Katamura K, Takeda H, Ohmura K, Yasumi T, Meguro TA, et al. Lipid A analogue, ONO-4007, inhibits IgE response and antigen-induced eosinophilic recruitment into airways in BALB/c mice. Int Arch Allergy Immunol. 2002;127: 217–25.
34. Cohn L, Elias JA, Chupp GL. Asthma: mechanisms of disease persistence and progression. Annu Rev Immunol. 2004;22:789–815.
35. Wills-Karp M. Immunologic basis of antigen-induced airway hyperresponsiveness. Annu Rev Immunol. 1999;17:255–81.
36. Wills-Karp M. Interleukin-13 in asthma pathogenesis. Curr Allergy Asthma Rep. 2004;4:123–31.
37. Williams T. The eosinophil enigma. J Clin Invest. 2004;113:507–9.
38. Laudanna C, Alon R. Right on the spot. Chemokine triggering of integrin-mediated arrest of rolling leukocytes. Thromb Haemost. 2006;95:5–11.
39. Voura EB, Billia F, Iscove NN, Hawley RG. Expression mapping of adhesion receptor genes during differentiation of individual hematopoietic precursors. Exp Hematol. 1997;25:1172–9.
40. Hynes RO. Integrins: bidirectional, allosteric signaling machines. Cell. 2002;110:673–87.
41. Hogg N, Laschinger M, Giles K, McDowall A. T-cell integrins: more than just sticking points. J Cell Sci. 2003;116:4695–705.
42. Gonzalez-Amaro R, Mittelbrunn M, Sanchez-Madrid F. Therapeutic anti-integrin (α4 and aL) monoclonal antibodies: two-edged swords? Immunology. 2005;116:289–96.
43. Lee S-H, Prince JE, Rais M, et al. Differential requirement for CD18 in T-helper effector homing. Nat Med. 2003;9:1281–6.

44. Lobb RR, Hemler ME. The pathophysiologic role of α4 integrins in vivo. J Clin Invest. 1994;94: 1722–8.
45. Nakajima H, Sano H, Nishimura T, Yoshida S, Iwamoto I. Role of vascular cell adhesion molecule 1/very late activation antigen 4 and intercellular adhesion molecule 1/lymphocyte function–associated antigen 1 interactions in antigen-induced eosinophil and T cell recruitment into the tissue. J Exp Med. 1994;179:1145–54.
46. Laberge S, Rabb H, Issekutz TB, Martin JG. Role of VLA-4 and LFA- 1 in allergen-induced airway hyperresponsiveness and lung inflammation in the rat. Am J Respir Crit Care Med. 1995;151: 822–9.
47. Chin JE, Hatfield CA, Winterrowd GE, et al. Airway recruitment of leukocytes in mice is dependent on α4-integrins and vascular cell adhesion molecule-1. Am J Physiol Lung Cell Mol Physiol. 1997;272:L219–29.
48. Henderson Jr WR, Chi EY, Albert RK, et al. Blockade of CD49d (α4 integrin) on intrapulmonary but not circulating leukocytes inhibits airway inflammation and hyperresponsiveness in a mouse model of asthma. J Clin Invest. 1997;100:3083–92.
49. Schneider T, Issekutz TB, Issekutz AC. The role of α4 (CD49d) and β2 (CD18) integrins in eosinophil and neutrophil migration to allergic lung inflammation in the brown Norway rat. Am J Respir Cell Mol Biol. 1999;20:448–57.
50. Borchers MT, Crosby J, Farmer S, et al. Blockade of CD49d inhibits allergic airway pathologies independent of effects on leukocyte recruitment. Am J Physiol Lung Cell Mol Physiol. 2001;280: L813–21.
51. Kanwar S, Smith CW, Shardonofsky FR, Burns AR. The role of Mac-1 (CD11b/CD18) in antigen-induced airway eosinophilia in mice. Am J Respir Cell Mol Biol. 2001;25:170–7.
52. Scott LM, Priestley GV, Papayannopoulou T. Deletion of α4 integrins from adult hematopoietic cells reveals roles in homeostasis, regeneration, and homing. Mol Cell Biol. 2003;23:9349–60.
53. Ulyanova T, Scott LM, Priestley GV, et al. VCAM-1 expression in adult hematopoietic and nonhematopoietic cells is controlled by tissue-inductive signals and reflects their developmental origin. Blood. 2005;106:86–94.
54. Priestley GV, Ulyanova T, Papayannopoulou T. Sustained alterations in biodistribution of stem/progenitor cells in Tie2Cre + α4f/f mice are hematopoietic cell autonomous. Blood. 2007;109:109–11.
55. Iwata A, Nishio K, Winn RK, Chi EY, Henderson Jr WR, Harlan JM. A broad-spectrum caspase inhibitor attenuates allergic airway inflammation in murine asthma model. J Immunol. 2003;170:3386–91.
56. Papayannopoulou T, Craddock C, Nakamoto B, Priestley GV, Wolf NS. The VLA4/VCAM-1 adhesion pathway defines contrasting mechanisms of lodgement of transplanted murine hemopoietic progenitors between bone marrow and spleen. Proc Natl Acad Sci U S A. 1995;92:9647–51.
57. Henderson Jr WR, Banerjee ER, Chi EY. Differential effects of (S)- and (R)-enantiomers of albuterol in a mouse asthma model. J Allergy Clin Immunol. 2005;116:332–40.
58. Miyahara N, Swanson B, Takeda K, et al. Effector CD8+ T cells mediate inflammation and airway hyper-responsiveness. Nat Med. 2004;10:865–9.
59. Humbles AA, Lloyd CM, McMillan SJ, et al. A critical role for eosinophils in allergic airways remodeling. Science. 2004;305:1776–9.
60. Lee JJ, Dimina D, Macias MP, et al. Defining a link with asthma in mice congenitally deficient in eosinophils. Science. 2004;305:1773–6.
61. Sehmi R, Baatjes AJ, Denburg JA. Hemopoietic progenitor cells and hemopoietic factors: potential targets for treatment of allergic inflammatory diseases. Curr Drug Targets Inflamm Allergy. 2003;2: 271–8.
62. Shang XZ, Armstrong J, Yang GY, et al. Regulation of antigen-specific versus by-stander IgE production after antigen sensitization. Cell Immunol. 2004;229:106–16.
63. Abonia JP, Hallgren J, Jones T, et al. a-4 integrins and VCAM-1, but not MAdCAM-1, are essential for recruitment of mast cell progenitors to the inflamed lung. Blood. 2006;108:1588–94.
64. Kramer MF, Jordan TR, Klemens C, et al. Factors contributing to nasal allergic late phase eosinophilia. Am J Otolaryngol. 2006;27:190–9.
65. Scharffetter-Kochanek K, Lu H, Norman K, et al. Spontaneous skin ulceration and defective T cell function in CD18 null mice. J Exp Med. 1998;188:119–31.
66. Koni PA, Joshi SK, Temann U-A, Olson D, Burkly L, Flavell RA. Conditional vascular cell adhesion molecule 1 deletion in mice: impaired lymphocyte migration to bone marrow. J Exp Med. 2001;193:741–54.
67. Barros MT, Acencio MMP, Garcia MLB, et al. BCG modulation of anaphylactic antibody response, airway inflammation and lung hyperreactivity in genetically selected mouse strains (selection IV-A). Life Sci. 2005;77:1480–92.
68. Snapper CM, Finkelman FD, Paul WE. Differential regulation of IgG1 and IgE synthesis by interleukin 4. J Exp Med. 1988;167:183–96.
69. Ohno H, Tsunemine S, Isa Y, Shimakawa M, Yamamura H. Oral administration of bifidobacterium bifidum G9-1 suppresses total and antigen specific immunoglobulin E production in mice. Biol Pharm Bull. 2005;28:1462–6.
70. Fish SC, Donaldson DD, Goldman SJ, Williams CMM, Kasaian MT. IgE generation and mast cell effector function in mice deficient in IL-4 and IL-13. J Immunol. 2005;174:7716–24.

71. Careau E, Bissonnette EY. Adoptive transfer of alveolar macrophages abrogates bronchial hyperresponsiveness. Am J Respir Cell Mol Biol. 2004;31:22–7.
72. Sato T, Tachibana K, Nojima Y, D'Avirro N, Morimoto C. Role of the VLA-4 molecule in T cell costimulation. Identification of the tyrosine phosphorylation pattern induced by the ligation of VLA-4. J Immunol. 1995;155:2938–47.
73. Nojima Y, Rothstein DM, Sugita K, Schlossman SF, Morimoto C. Ligation of VLA-4 on T cells stimulates tyrosine phosphorylation of a 105-kD protein. J Exp Med. 1992;175:1045–53.
74. Yamada A, Nojima Y, Sugita K, Dang NH, Schlossman SF, Morimoto C. Cross-linking of VLA/CD29 molecule has a co-mitogenic effect with anti-CD3 on CD4 cell activation in serum-free culture system. Eur J Immunol. 1991;21:319–25.
75. Nojima Y, Humphries MJ, Mould AP, et al. VLA-4 mediates CD3- dependent CD4+ T cell activation via the CS1 alternatively spliced domain of fibronectin. J Exp Med. 1990;172:1185–92.
76. Mittelbrunn M, Molina A, Escribese MM, et al. VLA-4 integrin concentrates at the peripheral supramolecular activation complex of the immune synapse and drives T helper 1 responses. Proc Natl Acad Sci U S A. 2004;101:11058–63.
77. Polte T, Foell J, Werner C, et al. CD137-mediated immunotherapy for allergic asthma. J Clin Invest. 2006;116:1025–36.
78. Jiang M-Z, Tsukahara H, Hayakawa K, et al. Effects of antioxidants and NO on TNF-α-induced adhesion molecule expression in human pulmonary microvascular endothelial cells. Respir Med. 2005;99: 580–91.
79. Wong CK, Wang CB, Li MLY, Ip WK, Tian YP, Lam CWK. Induction of adhesion molecules upon the interaction between eosinophils and bronchial epithelial cells: involvement of p38 MAPK and NF-kB. Int Immunopharmacol. 2006;6:1859–71.
80. Matsuno O, Miyazaki E, Nureki S, et al. Elevated soluble ADAM8 in bronchoalveolar lavage fluid in patients with eosinophilic pneumonia. Int Arch Allergy Immunol. 2006;142:285–90.
81. Wills-Karp M, Luyimbazi J, Xu X, et al. Interleukin-13: central mediator of allergic asthma. Science. 1998;282:2258–61.
82. Kuperman DA, Huang X, Koth LL, et al. Direct effects of interleukin- 13 on epithelial cells cause airway hyperreactivity and mucus overproduction in asthma. Nat Med. 2002;8:885–9.
83. Davidson E, Liub JJ, Sheikh A. The impact of ethnicity on asthma care. Prim Care Respir J. 2010;19(3):202–8.
84. Sharm P, Halayko AJ. Emerging molecular targets for the treatment of asthma. Indian J Biochem Biophys. 2009;46(6):447–60.
85. Broide DH, Sullivan S, Gifford T, Sriramarao P. Am J Respir Cell Mol Biol. 1998;18(2):218–25.
86. Takizawa H. Novel strategies for the treatment of asthma. Recent Pat Inflamm Allergy Drug Discov. 2007;1:13–9.
87. Czarnobilska E, Obtułowicz K. Eosinophil in allergic and non-allergic inflammation. Przegl Lek. 2005;62(12):1484–7.
88. Murphy DM, O'Byrne PM. Recent advances in the pathophysiology of asthma. Chest. 2010; 137(6):1417–26.
89. Woodside DG, Vanderslice P. Cell adhesion antagonists: therapeutic potential in asthma and chronic obstructive pulmonary disease. Biodrugs. 2008;22(2):85–100.
90. Erlandsen SL. Detection and spatial distribution of the beta 2 integrin (Mac- 1) and L-selectin (LECAM-1) adherence receptors on human neutrophils by high-resolution field emission SEM. J Histochem Cytochem. 1993;41:327–33.
91. Kuebler WM. Selectins revisited: the emerging role of platelets in inflammatory lung disease. J Clin Invest. 2006;116(12):3106–8.
92. Ray Banerjee E, Jiang Y, Henderson Jr WR, Scott LM, Papayannopoulou T. Alphα4 and beta2 integrins have non-overlapping roles in asthma development, but for optimal allergen sensitization only alphα4 is critical. Exp Hematol. 2007;35(4): 605–1.
93. Ray Banerjee E, Jiang Y, Henderson Jr WR, Latchman YL, Papayannopoulou T. Absence of α4 but not β2 integrins restrains the development of chronic allergic asthma using mouse genetic models. Exp Hematol. 2009;37:715–27.
94. Bevilacqua MP. Setectins. J Clin Invest. 1993; 91:379.
95. Curtis JL, Sonstein J, Craig RA, Todt JC, Knibbs RN, Polak T, et al. Subset-specific reductions in lung lymphocyte accumulation following intratracheal antigen challenge in endothelial selectin-deficient mice. J Immunol. 2002;169:2570–9.
96. Harlan JM, Liu DY. Adhesion: its role in inflammatory disease. New York: W.H. Freeman; 1992. p. 85–115.
97. Crockett-Torabi E. Selectins and mechanisms of signal transduction. J Leukocyte Biol. 1998;63(1): 1–14.
98. Harlan JM. Leukocyte adhesion deficiency syndrome: insights into the molecular basis of leukocyte emigration. Clin Immunol Immunopathol. 1993;67(3 Pt 2):S16–24.
99. De Sanctis GT et al. Inhibition of pulmonary eosinophilia in P-selectin and ICAM-1-deficient mice. J Appl Physiol. 1997;83(3):681.
100. Lukacs NW et al. Reduction of allergic airway responses in P-selectin deficient mice. J Immunol. 2002;169(4):2120–5.
101. Zarbock A, Singbartl K, Ley K. Complete reversal of acid-induced acute lung injury by blocking of platelet-neutrophil aggregation. J Clin Invest. 2006;116(12):3211–19.
102. Engelberts I, Samyo SK, Leeuwenberg JF, van der Linden CJ, Buurman WA. A role for ELAM-1 in the pathogenesis of MOF during septic shock. J Surg Res. 1992;53(2):136–44.

103. Eppihimer MJ, Russell J, Langley R, Gerritsen M, Granger DN. Role of tumor necrosis factor and interferon gamma in endotoxin-induced E-selectin expression. Shock. 1999;11(2):93–7.
104. Keelan ET, Licence ST, Peters AM, Binns RM, Haskard DO. Characterization of E-selectin expression in vivo with use of a radiolabeled monoclonal antibody. Am J Physiol. 1994;266(1 Pt 2):H278–90.
105. Redl H, Dinges HP, Buurman WA, van der Linden CJ, Pober JS, Cotran RS, et al. Expression of endothelial leukocyte adhesion molecule-1 in septic but not traumatic/hypovolemic shock in the baboon. Am J Pathol. 1991;139(2):461–6.
106. Steinhoff G, Behrend M, Schrader B, Pichlmayr R. Intercellular immune adhesion molecules in human liver transplants: overview on expression patterns of leukocyte receptor and ligand molecules. Hepatology. 1993;18(2):440–53.
107. Jutila MA, Walcheck B, Bargatze R, Palecanda A. Measurement of neutrophil adhesion under conditions mimicking blood flow. Methods Mol Biol. 2007;412:239–56.
108. Araki M, Araki K, Miyazaki Y, Iwamoto M, Izui S, Yamamura K, et al. E-selectin binding promotes neutrophil activation in vivo in E-selectin transgenic mice. Biochem Biophys Res Commun. 1996;224(3):825–30.
109. Kulidjian AA, Issekutz AC, Issekutz TB. Differential role of E-selectin and P-selectin in T lymphocyte migration to cutaneous inflammatory reactions induced by cytokines. Int Immunol. 2002;14(7): 751–60.
110. Robinson SD, Frenette PS, Rayburn H, Cummiskey M, Ullman-Culleré M, Wagner DD, et al. Multiple, targeted deficiencies in selectins reveal a predominant role for P-selectin in leukocyte recruitment. Proc Nat Acad Sci USA. 1999;96(20):11452–7.
111. Kelly Margaret BJ. Current reviews of allergy and clinical immunology. Immunol Today. 1999;20:545–50.
112. Schleimer RP. The role of adhesion molecules in allergic inflammation and their suitability as targets of antiallergic therapy. Clin Exp Allergy. 1998;3: 15–23.
113. DD, W Ciba Found Symp. 1995:2–10, discussion 10–16, 77–8.
114. Tang MLK, Fiscus LC. Important roles for L-selectin and ICAM-1 in the development of allergic airway inflammation in asthma. Pulmon Pharmacol Ther. 2001;14:203–10.
115. Bowden RA, Ding ZM, Donnachie EM, Petersen TK, Michael LH, Ballantyne CM, et al. Role of alphaα4 integrin and VCAM-1 in CD18- independent neutrophil migration across mouse cardiac endothelium. Circ Res. 2002;90(5):562–9.
116. Pendl GG, Robert C, Steinert M, Thanos R, Eytner R, Borges E, et al. Immature mouse dendritic cells enter inflamed tissue, a process that requires E- and P-selectin, but not P-selectin glycoprotein ligand 1. Blood. 2002;99(3):946–56.
117. Haugen TS, Skjonsberg O, Nakstad B, Lyberg T. Modulation of adhesion molecule profiles on alveolar macrophages and blood leukocytes. Respiration. 1999;66(6):528–37.
118. Gonzalo JA, Lloyd CM, Kremer L, Finger E, Martinez-A C, Siegelman MH, et al. Eosinophil recruitment to the lung in a murine model of allergic inflammation. The role of T cells, chemokines, and adhesion receptors. J Clin Invest. 1996;98(10):2332–45.
119. Umetsu DT, McIntire JJ, Akbari O, Macaubas C, DeKruyff RH. Asthma: an epidemic of dysregulated immunity. Nat Immunol. 2002;3:715–20.
120. Peroz Novo CA, Jedrzejczak-Czechowicz M, Lewandowska-Polak A, Claeys C, et al. T cell inflammatory response, Foxp3 and TNFRS18-L regulation of peripheral blood mononuclear cells from patients with nasal polypsasthma after staphylococcal superantigen stimulation. Clin Exp Allergy. 2010;40:1323–32.
121. Harris JF, Fischer MJ, Hotchkiss JR, Monia BP, Randell SH, Harkema JR, et al. Bcl-2 sustains increased mucous and epithelial cell numbers in metaplastic airway epithelium. Am J Respir Crit Care Med. 2005;171(7):764–72.
122. Randolph DA, Carruthers CJL, Szabo SJ, Murphy KM, Chaplin DD. Modulation of airway inflammation by passive transfer of allergen- specific Th1 and Th2 cells in a mouse model of asthma. J Immunol. 1999;162:2375–83.
123. Rossi B, Constantin G. Anti-selectin therapy for the treatment of inflammatory diseases. Inflamm Allergy Drug Targets. 2008;7:85–93.
124. Groemping Y, Rittinger K. Activation and assembly of the NADPH oxidase: a structural perspective. Biochem J. 2005;386:401–16.
125. Henriet SS, Hermans PW, Verweij PE, Simonetti E, Holland SM, Sugui JA, et al. Human leucocytes kill Aspergillus nidulans by ROS-independent mechanisms. Infect Immun. 2010;79:767–73.
126. Bylund J, Brown KL, Movitz C, Dahlgren C, Karlsson A. Intracellular generation of superoxide by the phagocyte NADPH oxidase: how, where, and what for? Free Radic Biol Med. 2010;49:1834–45.
127. Kumar S, Patel S, Jyoti A, Keshari RS, Verma A, Barthwal MK, et al. Nitric oxide-mediated augmentation of neutrophil reactive oxygen and nitrogen species formation: critical use of probes. Cytometry A. 2010 Nov;77(11):1038–48.
128. De Ravin SS, Zarember KA, Long-Priel D, Chan KC, Fox SD, Gallin JI, et al. Tryptophan/kynurenine metabolism in human leukocytes is independent of superoxide and is fully maintained in chronic granulomatous disease. Blood. 2010 Sep 9;116(10): 1755–60.
129. Chan EC, Dusting GJ, Guo N, Peshavariya HM, Taylor CJ, Dilley R, et al. Prostacyclin receptor suppresses cardiac fibrosis: role of CREB phosphorylation. J Mol Cell Cardiol. 2010 Aug;49(2): 176–85.

130. Leverence JT, Medhora M, Konduri GG, Sampath V. Lipopolysaccharide-induced cytokine expression in alveolar epithelial cells: Role of PKCζ-mediated p47phox phosphorylation. Chem Biol Interact. 2010;189:72–81.
131. Kim Y, Zhou M, Moy S, Morales J, Cunningham MA, Joachimiak A. High-resolution structure of the nitrile reductase QueF combined with molecular simulations provide insight into enzyme mechanism. J Mol Biol. 2010 Nov 19;404(1):127–37.
132. Santilli G, Almarza E, Brendel C, Choi U, Beilin C, Blundell MP, et al. Biochemical correction of X-CGD by a novel chimeric promoter regulating high levels of transgene expression in myeloid cells. Mol Ther. 2010;19:122–32.
133. Kassim SY, Fu X, Liles WC, Shapiro SD, Parks WC, Heinecke JW. NADPH oxidase restrains the matrix metalloproteinase activity of macrophages. J Biol Chem. 2005 Aug 26;280(34):30201–5.
134. Lanone S, Zheng T, Zhu Z, Liu W, Lee CG, Ma B, et al. Overlapping and enzyme-specific contributions of matrix metalloproteinases-9 and −12 in IL-13–induced inflammation and remodelling. J Clin Invest. 2002;110(4):463–74.
135. Kassim SY. NADPH oxidase restrains the matrix metalloproteinase activity of macrophages. JBC. 2005;280(34):30201–5.
136. Nurieva RI, Xoi Moui Mai, Forbush K, Bevan MJ, Chen Dong. B7h is required for T cell activation, differentiation, and effector function. PNAS. 2003;100:14163.
137. Wang S, Zhu G, Chapoval AI, Dong H, Tamada K, Ni J, et al. Costimulation of T cells by B7-H2, a B7-like molecule that binds ICOS. Blood. 2000;96(8):2808–281.
138. Suh WK, Tafuri A, Berg-Brown NN, Shahinian A, Plyte S, Duncan GS, et al. The inducible costimulator plays the major costimulatory role in humoral immune responses in the absence of CD28. J Immunol. 2004 May 15;172(10):5917–23.
139. Hutloff A, Dittrich AM, Beier KC, Eljaschewitsch B, Kraft R, Anagnostopoulos I, et al. ICOS is an inducible T-cell co-stimulator structurally and functionally related to CD28. Nature. 1999 Jan 21;397(6716):263–6.
140. Yoshinaga SK. T-cell co-stimulation through B7RP-1 and ICOS. Nature. 1999 Dec 16;402(6763): 827–32.
141. Qian X, Agematsu K, Freeman GJ, Tagawa Y, Sugane K, Hayashi T. The ICOS-ligand B7-H2, expressed on human type II alveolar epithelial cells, plays a role in the pulmonary host defense system. Eur J Immunol. 2006 Apr;36(4):906–18.
142. Schuyler M, Gott K, Edwards B. Th1 cells that adoptively transfer experimental hypersensitivity pneumonitis are activated memory cells. Cell Immunol. 1999;177(6):377–38.
143. Gudmundsson G, Hunninghake GW. Interferon-gamma is necessary for the expression of hypersensitivity pneumonitis. J Clin Invest. 1997;99: 2386–90.
144. Gudmundsson G, Monick MM, Hunninghake GW. IL-12 modulates expression of hypersensitivity pneumonitis. J Immunol. 1998;161:991.
145. Meyer F, Ramanujam KS, Gobert AP, James SP, Wilson KT. Cycloxygenase-2 activation suppresses Th1 polarization in response to *Helicobacter pylori*. J Immunol. 2003;171:3931.
146. Lenschow DJ, Walunas TL, Bluestone JA. CD28/B7 system of T cell costimulation. Annu Rev Immunol. 1996;14:233.
147. Linsley PS, Wallace PM, Johnsom J, Gibson MG, Greene JL, Ledbetter JA, et al. Immunosuppression in-vivo by a soluble form of the CTLA4 T cell activation molecule. Science. 1992;257:792.
148. Thompson CB. Distinct roles for the costimulatory ligands B7-1 and B7-2 in T helper cell differentiation? Cell. 1995 Jun 30;81(7):979–82.
149. Schweitzer AN, Sharpe AH. Studies using antigen-presenting cells lacking expression of both B7-1 (CD80) and B7-2 (CD86) show distinct requirements for B7 molecules during priming versus restimulation of Th2 but not Th1 cytokine production. J Immunol. 1998;161:2762.
150. Chelen CJ, Fang Y, Freeman GJ, Secrist H, Marshall JD, Hwang PT, et al. Human alveolar macrophages present antigen ineffectively due to defective expression of B7 costimulatory cell surface molecules. J Clin Invest. 1995;95(3):1415–21.

Published in

Henderson WR, Banerjee ER, Chi EY. Differential effects of (S)- and (R)- enantiomers of albuterol in mouse asthma model. J Allergy Clin Immunol. 2005;116:332–40.

Banerjee ER. Triple selectin knockout (ELP–/–) mice fail to develop OVA-induced acute asthma phenotype. J Inflamm. 2011;8:19. doi:10.1186/1476-9255-8-19.

Banerjee ER, Henderson Jr WR. NADPH oxidase has a regulatory role in acute allergic asthma. J Adv Lab Res Biol. 2011;2(3):103–20.

Banerjee ER, Henderson Jr WR. Defining the molecular role of gp91phox in the manifestation of acute allergic asthma using a preclinical murine model. Clin Mol Allergy. 2012;10(1):2–16.

Banerjee ER, Henderson Jr WR. Role of T cells in a gp91phox knockout murine model of acute allergic asthma. Allergy Asthma Clin Immunol. 2013;9(1):6. [Epub ahead of print] PubMed PMID: 23390895.

Role of Integrins α4 and β2 Onset and Development of Chronic Allergic Asthma in Mice

4

Abstract

Objective. Chronic asthma is characterized by an ongoing recruitment of inflammatory cells and airway hyperresponsiveness leading to structural airway remodeling. Although α4b1 and β2 integrins regulate leukocyte migration in inflammatory diseases and play decisive roles in acute asthma, their role has not been explored under the chronic asthma setting. To extend our earlier studies with α4Δ/Δ and β2L/L mice, which showed that both α4 and β2 integrins have nonredundant regulatory roles in acute ovalbumin (OVA)-induced asthma, we explored to what extent these molecular pathways control development of structural airway remodeling in chronic asthma.

Materials and Methods. Control, α4Δ/Δ, and β2L/L mouse groups, sensitized by intraperitoneal OVA as allergen, received intratracheal OVA periodically over days 8–55 to induce a chronic asthma phenotype. Post-OVA assessment of inflammation and pulmonary function (airway hyperresponsiveness), together with airway modeling measured by goblet cell metaplasia, collagen content of lung, and transforming growth factor b1 expression in lung homogenates were evaluated.

Results. In contrast to control and β2L/L mice, α4Δ/Δ mice failed to develop and maintain the composite chronic asthma phenotype evaluated as mentioned, and subepithelial collagen content was comparable to baseline. These data indicate that β2 integrins, although required for inflammatory migration in acute asthma, are dispensable for structural remodeling in chronic asthma.

Conclusion. α4 integrins appear to have a regulatory role in directing transforming growth factor b-induced collagen deposition and structural alterations in lung architecture likely through interactions of Th2 cells, eosinophils, or mast cells with endothelium, resident airway cells, and/or extracellular matrix.

E.R. Banerjee, *Perspectives in Inflammation Biology*,
DOI 10.1007/978-81-322-1578-3_4, © Springer India 2014

Background and Objective of the Research Undertaken

In acute asthma, dysregulated immunity triggers a Th2 response by antigen-presenting cells and Th2-derived cytokines, especially interleukin-4 (IL-4) and IL-13, promoting B-cell differentiation into immunoglobulin (Ig) E-sequestering plasma cells. Cross-linking of IgE receptors on mast cells releases histamines, prostaglandins, thromboxane, and leukotrienes, leading to bronchoconstriction, vasodilation, and mucus secretion [1]. Thus, a cascade of interactions between cells and soluble molecules in the airways result in bronchial mucosal inflammation and airway hyperresponsiveness (AHR) [2]. In chronic allergic asthma, there is continuous recruitment of Th2 as well as inflammatory cells in the lung and airways. These cells and their secreted products elicit structural changes in resident airway cells, including epithelial desquamation, goblet cell metaplasia, mucus hypersecretion, and thickening of submucosa, manifested as bronchoconstriction and AHR [3, 4]. Prominent in the remodeling process is the thickening of the airway wall with development of subepithelial fibrosis from deposition of extracellular matrix proteins, such as collagen, laminin, fibronectin, and tenascin in the lamina reticularis beneath the basement membrane [5–11]. Despite the fact that the histologic features of airway remodeling in chronic asthma have been well characterized, the immunologic and inflammatory mechanisms that maintain or enhance remodeling are incompletely understood. Although mouse models of asthma do not completely reproduce all the hallmarks of human disease and several pathophysiologic responses in mice have been of limited value in humans, these models have provided important insights into the pathophysiology of asthma and have been used for testing new treatments of allergic asthma [12]. Using mouse models, it was found that leukocyte migration into lung is an important early event in the pathogenesis of asthma, and it is mediated by a series of adhesive interactions between leukocytes and airway cells for which integrins (α2 and b1) have been found to be critical participants [13–18]. While CD18 (β2 integrin) null mice have been used to investigate the role of CD18 in allergic asthma [19], studies on α4 integrins have been previously limited to those using monoclonal antibodies or other inhibitors of α4 integrin [13–15, 17, 18, 20, 21]. Our recent studies with conditionally ablated α4 knockout mice tested in parallel with CD18−/− mice showed that, while β2 integrins control inflammatory migration in the airways, α4 integrins subvert the onset of acute asthma by curtailing the initial sensitization process, as well as by preventing cross talk between inflammatory leukocytes and their interaction with the endothelium and lung stroma [22]. Because chronic rather than acute asthma appears to be more relevant to human disease [23], it was important to explore the involvement of these two types of integrins in the chronic setting of allergen challenge. Thus, using a repeated-challenge protocol in a more chronic setting, we assessed, in these genetic mouse models, changes associated with structural remodeling of the airways to gain further insight into the contribution of α4 and β2 integrins to the airway remodeling in chronic allergic asthma. Our data uncover novel information about the differential contribution of β2 versus α4 integrins in the composite phenotype of chronic asthma development and contribute to the understanding of mechanisms by which different cell subsets and molecular pathways participate in the pathophysiology and histopathology of chronic asthma.

Results in a Nutshell

In contrast to control and β2L/L mice, α4Δ/Δ mice failed to develop and maintain the composite chronic asthma phenotype evaluated as mentioned, and subepithelial collagen content was comparable to baseline. These data indicate that β2 integrins, although required for inflammatory migration in acute asthma, are dispensable for structural remodeling in chronic asthma.

Detailed Results

α4Δ/Δ Mice Fail to Develop AHR to Chronic Airway Challenge by Allergen

To determine pulmonary function in the OVA-treated mice, we measured AHR to increasing doses of methacholine by noninvasive whole-body plethysmography (Penh) (Fig. 4.1b). While both post-OVA control and CD18−/− mice show significantly increased Penh value over saline-treated controls at 40 and 100 mg/mL methacholine, OVA-treated α4Δ/Δ mice have a Penh value similar to baseline values from untreated or saline-treated mice. Using the same genetic models (α4Δ/Δ, CD18−/−), we have previously shown [22] that measurements of airway resistance by invasive plethysmography were in full agreement with responses by noninvasive whole-body plethysmography, as measured here.

Migration of Leukocytes from Circulation to Lung and to Airways

Total number of lymphoid and myeloid cells present in BM, peripheral blood circulation, spleen, or in BALf and lung parenchyma was assessed through measurements of total nucleated cell counts (Fig. 4.1c, d), and their differentials were determined by FACS based on specific antigen expression and by cell morphology from smears (Fig. 4.2 and Table 4.1). Both total cell numbers and differentials varied among the animal groups and tissues studied. It is worth noting that differences among the groups also exist at baseline (i.e., increase in leukocytes in peripheral blood and spleen), as previous studies with these mice indicate [19, 24]. In BM and peripheral blood, the total number of different leukocyte subsets was higher post-OVA in CD18−/− mice compared to controls. α4Δ/Δ mice also had higher levels than controls in BM, with one exception, i.e., eosinophils. In lung parenchyma, CD18−/− mice again had the highest numbers of leukocytes, reflecting both recruitment and accumulation in the lung. In BALf, all leukocyte subsets were low in α4Δ/Δ mice, whereas only lymphocytes and eosinophils were reduced in CD18−/− mice (Fig. 4.2) compared to controls. It is of interest that, as in acute asthma, eosinophils were not increased in BALf of both CD18−/− and α4Δ/Δ mice, suggesting that for the interstitial airway migration of these cells, β2 and α4 are both important and nonredundant. However, chronic accumulation of eosinophils in lung parenchyma is α4- and β2-integrin independent. Nevertheless, despite minimal eosinophil presence in BALf and increased presence of eosinophils in the lungs of both genetic models, AHR was induced in CD18−/− mice only, but not in α4Δ/Δ mice.

Inflammation and Fibrosis in the Lungs in Response to Chronic OVA Challenge

To assess the degree of inflammation and structural fibrotic remodeling in response to chronic allergen challenge, paraffin-embedded lung sections from untreated and OVA-treated mice were stained with hematoxylin and eosin to detect inflammatory cell migration into the lung, Alcian blue to detect goblet cell metaplasia and mucus secretion in the luminal spaces of the lung, and Masson's trichrome and Martius scarlet blue to assess changes primarily in lung collagen content (Fig. 4.3). While OVA-treated control and CD18−/− lungs show considerable inflammatory changes and increased deposition of collagen, indicating subepithelial fibrosis compared to aluminum sulfate-treated mice, in OVA-treated α4Δ/Δ mice no significant changes from non-OVA-treated control mice were noted. Mononuclear cell accumulation in the lungs of α4Δ/Δ mice was frequently in a nodular pattern within the parenchyma (Fig. 4.3, right panel), unlike the other models.

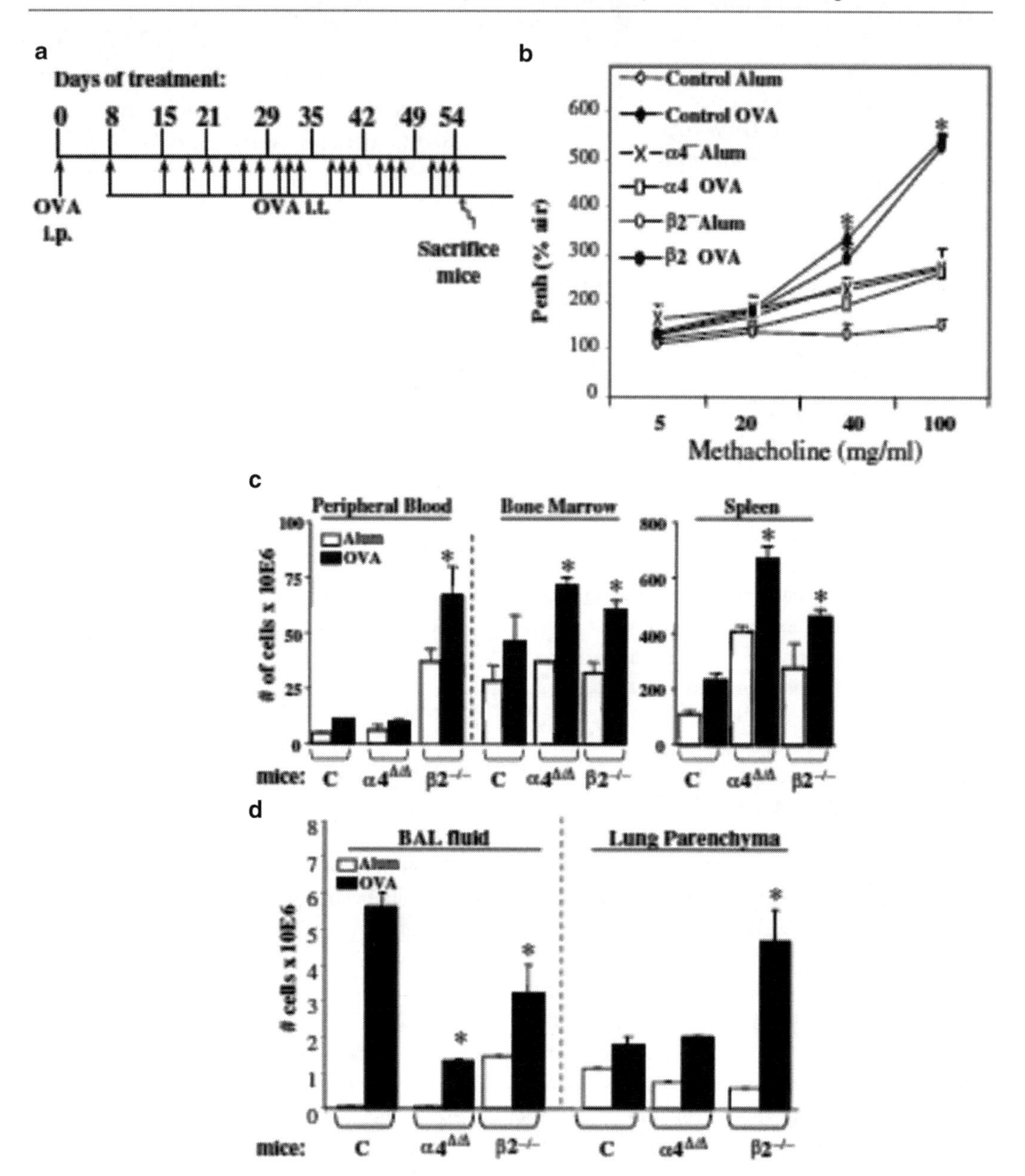

Fig. 4.1 (**a**) Diagram illustrating the experimental design for chronic asthma development. *i.p.* intraperitoneal injection, *i.t.* intratracheal instillation. (**b**) Lung function testing using whole-body plethysmography. Measurement of airway hyperresponsiveness at 24 h after the last ovalbumin (OVA) challenge on day 54. The degree of bronchoconstriction to increasing doses of aerosolized methacholine was expressed as Penh (percentage of air as control). $*p < 0.01$ compared to baseline values in saline-treated control mice, $n = 8$ mice per group. Control mice, α4-deficient mice, and β2-deficient mice were treated with aluminum sulfate (alum) and OVA. Measurements of airway resistance by invasive plethysmography [22] were found to be in full agreement with Penh data. (**c**) Numbers of lymphoid and myeloid cells in blood (n/mL), bone marrow (n/femur), and spleen, and (**d**) total numbers of cells in bronchoalveolar lavage (BAL) fluid and lung parenchyma from the two genetic models and controls treated with alum or OVA. Data are from two independent experiments expressed as average ± standard error of the mean. $n = 8$ mice per group

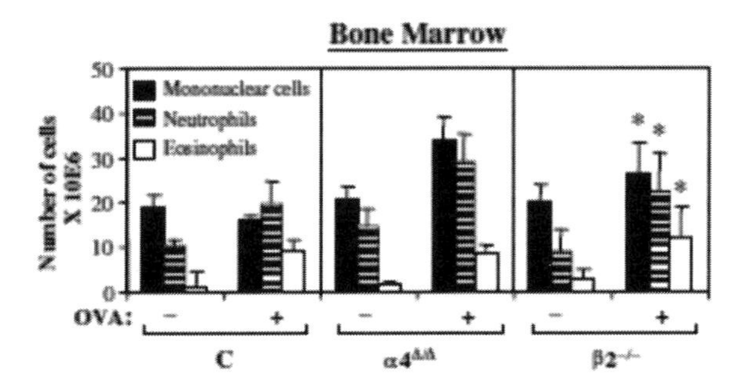

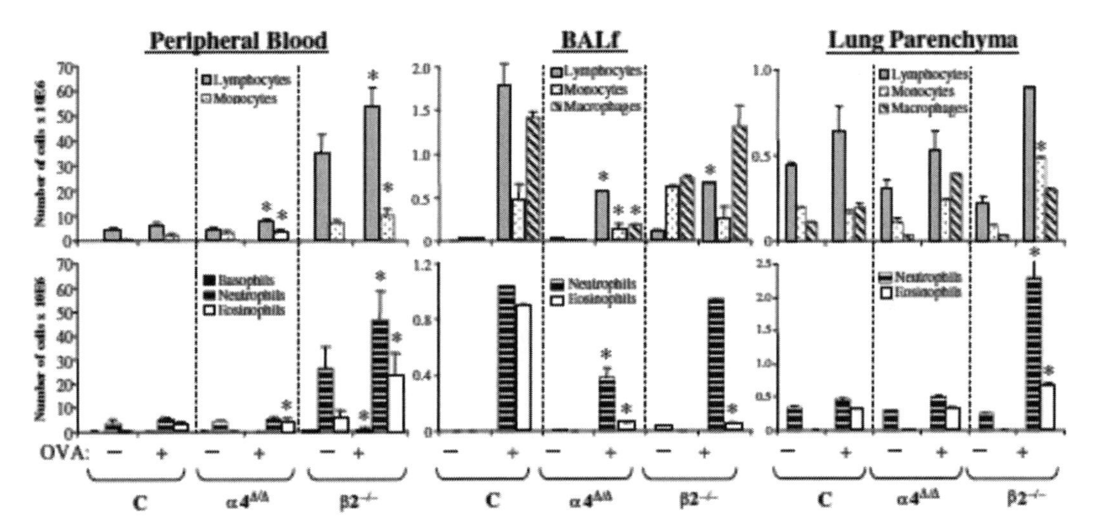

Fig. 4.2 Total numbers of all leukocytes in bone marrow, peripheral blood, bronchoalveolar lavage fluid (BALf), and lung parenchyma in control, α4-deficient (α4Δ/Δ), and $\beta2^{-/-}$ mice pre- and post-chronic ovalbumin (OVA) treatment. Total numbers (×10E6) of cells were counted in a Beckman Coulter Counter and differentials (neutrophils, eosinophils, monocytes, etc.) were assessed by fluorescein-activated cell sorting (specific antigen expression) and morphology in smears. $*p < 0.01$ from control values post-OVA, $n = 8$ mice per group, *C* control mice

Th2/Th1 Cytokines in BALf and Plasma and IgE and IgG1 Levels in Plasma

Levels of Th1 and Th2 cytokines and eotaxin were measured in plasma and BALf as described in "Materials and Methods." All Th2 cytokines (i.e., IL-4, IL-5, IL-13), eotaxin, and TNFa levels were low in BALf (Fig. 4.4a, left panels) of α4Δ/Δ mice compared to both control and β2 mice post-OVA, whereas in plasma (Fig. 4.4a, right panels), IL-13 levels were similar to other groups. CD18−/− mice developed high levels of cytokines in circulation, like control mice, but in BALf (Fig. 4.4), only TNF and eotaxin were significantly lower than controls (see Table 4.2), commensurate with the lower eosinophil levels (Fig. 4.2). Furthermore, OVA-sensitization responses were largely attenuated in α4Δ/Δ mice in contrast to the other two groups, as indicated by levels of OVA-specific IgE in plasma (Fig. 4.4c), corroborating previous findings in acute asthma [22].

Soluble VCAM-1 in BALf and Plasma and VCAM-1 Expression in the Lung

Because increases in cytokine levels (e.g., IL-4, IL-13, TNF-α) upregulate expression of VCAM-1 and increased VCAM-1 is a hallmark of allergic asthma phenotype, we next investigated whether the presence or absence of asthma phenotype

Table 4.1 Total numbers of nucleated cells (×10E6) in bone marrow, blood airways, and lung (data for Fig. 4.2)

BM/femur	Mononuclear			Neutrophils	Eosinophils	Total
WT + alum	19.087 ± 2.56			10.459 ± 1.08	1.229 ± 3.38	30.772 ± 2.36
WT + OVA	16.038 ± 1.07			19.718 ± 4.96	9.394 ± 2.28	45.15 ± 1.15
$\alpha4^{\Delta/\Delta}$ + alum	20.371 ± 3.39			14.714 ± 3.97	1.914 ± 0.46	37 ± 5.91
$\alpha4^{\Delta/\Delta}$ + OVA	33.734 ± 5.51			29.209 ± 5.93	8.557 ± 1.75	715 ± 7.94
$\beta2^{-/-}$ + alum	20.077 ± 3.78			8.874 ± 5.06	3.189 ± 11.86	32.14 ± 4.98
$\beta2^{-/-}$ + OVA	**26.180 ± 7.05**			**22.637 ± 8.53**	**12.027 ± 6.97**	**60.845 ± 3.08**
PB/2 ml	Lymphocytes	Monocytes	Basophils			
WT + alum	4.231 ± 1.12	0.529 ± 0.07	0.132 ± 0.07	3.487 ± 1.96	0.578 ± 0.03	8.95 ± 0.35
WT + OVA	6.616 ± 0.86	2.214 ± 0.67	0.117 ± 0.06	5.062 ± 0.75	3.159 ± 0.86	17.167 ± 2.32
$\alpha4^{\Delta/\Delta}$ + alum	4.378 ± 0.94	3.079 ± 0.86	0.000 ± 0.00	4.247 ± 0.27	0.161 ± 0.15	11.864 ± 1.07
$\alpha4^{\Delta/\Delta}$ + OVA	**7.594 ± 1.07**	**3.379 ± 0.83**	0.062 ± 0.07	5.050 ± 1.06	**3.978 ± 1.96**	**20.064 ± 8.65**
$\beta2^{-/-}$ + alum	34.425 ± 7.85	6.799 ± 1.58	0.258 ± 0.01	26.766 ± 8.86	5.852 ± 2.96	74.1 ± 15.61
$\beta2^{-/-}$ + OVA	**53.328 ± 7.97**	**9.929 ± 2.97**	**0.723 ± 0.97**	**46.658 ± 11.96**	**23.475 ± 9.28**	**133.814 ± 64.96**
BALf/2lungs			Macrophages			
WT + alum	0.003 ± 0.0001	0.030 ± 0.0007	0.023 ± 0.005	0.003 ± 0.0001	0.000 ± 0	0.06 ± 0.003
WT + OVA	1.778 ± 0.265	0.466 ± 0.18	1.409 ± 0.067	1.040 ± 0.003	0.908 ± 0.002	5.6 ± 0.48
$\alpha4^{\Delta/\Delta}$ + alum	0.029 ± 0.0074	0.017 ± 018	0.009 ± 0.0001	0.005 ± 0.0001	0.000 ± 0	0.06 ± 0.001
$\alpha4^{\Delta/\Delta}$ + OVA	**0.566 ± 0.0076**	**0.130 ± 0.056**	**0.176 ± 0.007**	**0.379 ± 0.072**	**0.062 ± 0.001**	**1.314 ± 0.013**
$\beta2^{-/-}$ + alum	0.107 ± 0.007	0.594 ± 0.024	0.715 ± 0.023	0.041 ± 0.001	0.000 ± 0	1.456 ± 0.039
$\beta2^{-/-}$ + OVA	**0.658 ± 0.003**	0.258 ± 0.131	1.312 ± 0.244	0.941 ± 0.002	**0.050 ± 0.002**	3.22 ± 0.78
LP/lung						
WT + alum	0.442 ± 0.012	0.192 ± 0.007	0.106 ± 0.004	0.356 ± 0.003	0.000 ± 0	1.096 ± 0.023
WT + OVA	0.636 ± 0.154	0.158 ± 0.023	0.193 ± 0.023	0.476 ± 0.001	0.315 ± 0.001	1.777 ± 0.164
$\alpha4^{\Delta/\Delta}$ + alum	0.304 ± 0.054	0.108 ± 0.024	0.031 ± 0.0001	0.304 ± 0.001	0.000 ± 0	0.747 ± 0.003
$\alpha4^{\Delta/\Delta}$ + OVA	0.525 ± 0.12	0.241 ± 0.001	0.388 ± 0.005	0.506 ± 0.002	0.340 ± 0.002	2 ± 0.045
$\beta2^{-/-}$ + alum	0.226 ± 0.03	0.094 ± 0.0002	0.033 ± 0.0001	0.247 ± 0.007	0.000 ± 0	0.6 ± 0.002
$\beta2^{-/-}$ + OVA	0.894 ± 0.002	**0.486 ± 0.003**	0.301 ± 0.002	**2.287 ± 0.56**	**0.692 ± 0.002**	**4.66 ± 0.893**

Numbers in bold indicate $p < 0.01$
Alum aluminum sulfate, *BM* bone marrow, *LP* lung parenchyma, *OVA* ovalbumin, *PB* peripheral blood, *WT* Wild Type

affected the levels of soluble VCAM-1 in plasma and BALf or its expression in the lung in the two genetic models compared to controls. Soluble VCAM-1 levels post-OVA in both plasma and BALf were increased in control and CD18−/− mice, but in α4Δ/Δ mice were quite low by comparison (Fig. 4.5a, b). Histochemical evidence of VCAM-1 protein in the lung was also increased in response to OVA in both control and CD18−/− lung (Fig. 4.6c, left panel), similar to changes seen in acute asthma [22]. However, in α4Δ/Δ lung post-OVA expression of VCAM-1 (Fig. 4.6c, right panel) was almost unchanged from baseline and lower than what was observed in the other two groups post-OVA. Therefore, although reduced levels of relevant cytokines could be responsible for decreased upregulation of VCAM-1 in α4Δ/Δ mice, other reasons, e.g., unresponsive endothelium in α4Δ/Δ mice, cannot be excluded.

TGF-β1 and Soluble Collagen in BALf

TGF-β1 has been thought to be instrumental in fibrosis development and consequent collagen deposition in the lung. Therefore, because subendothelial collagen deposition was present in control and CD18−/− mice by immunohistochemistry (Fig. 4.3), we measured TGF-b1 in BALf (Fig. 4.5c) and soluble collagen levels in the lung (Fig. 4.5d) of these mice. Significant increases in TGF-b1 and soluble collagen from aluminum sulfate-treated mice were only seen in control and

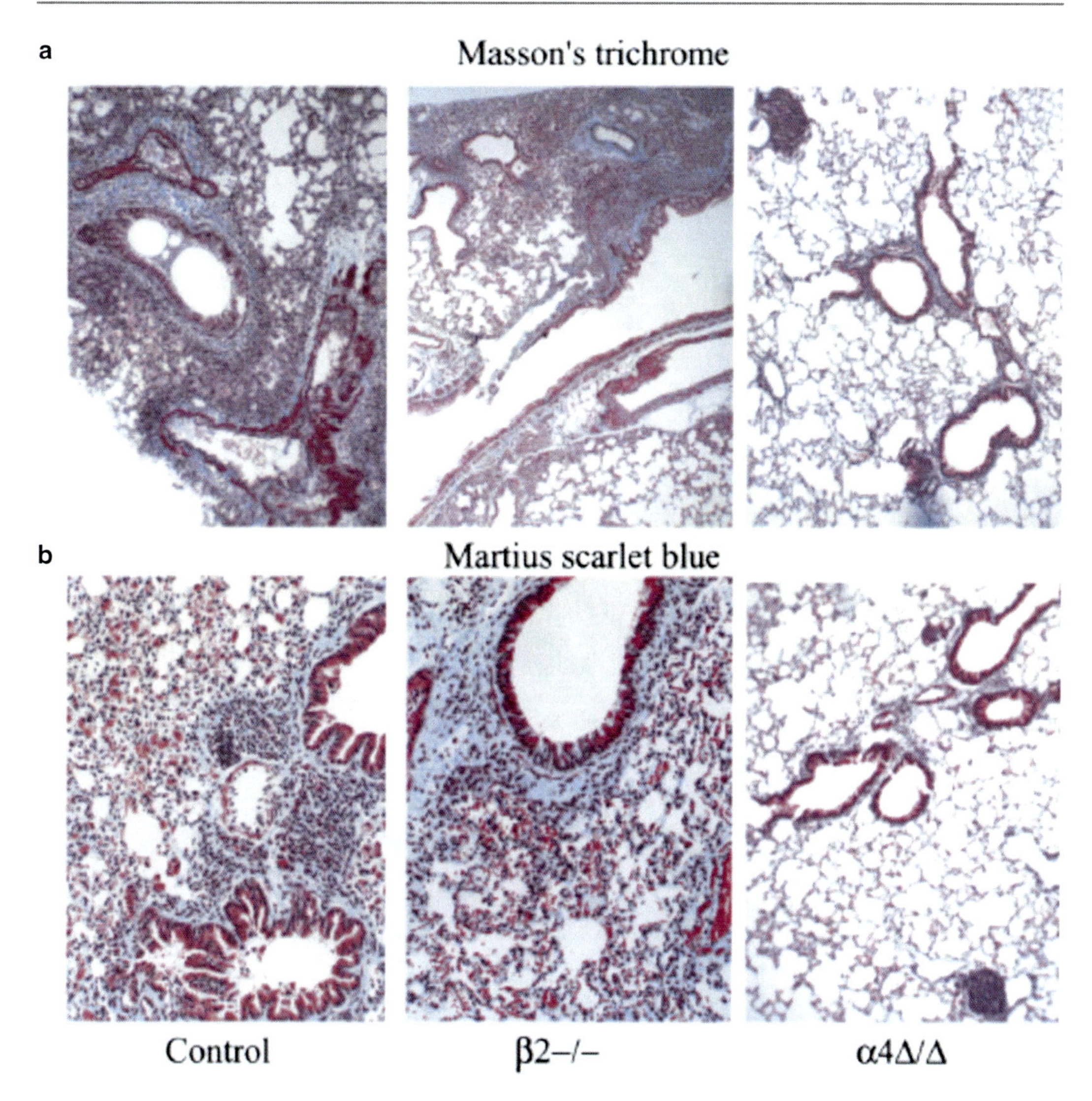

Fig. 4.3 Paraffin sections of lungs from ovalbumin-treated mice were stained with (**a**) Masson's trichrome stain revealing the collagen depositions around the airways and (**b**) Martius scarlet blue for collagen and fibrin deposition in interstitial spaces of the parenchyma. Adjustments for brightness, contrast, and color balance using Adobe Photoshop were made in order to match in all six photographs

CD18–/– mice post-OVA. In contrast, in α4Δ/Δ mice, levels were very low consistent also with immunohistochemical documentation of TGF-b1 in lung sections (Fig. 4.6a, right panel).

Discussion

Although a body of literature is available describing the development of acute asthma in animal models, data comparing asthma and airway remodeling in both acute and chronic asthma settings by employing the same genetic mouse models are rare. As chronic asthma is considered more relevant to the human disease, molecular pathways involved in its development are important for designing targeted therapies [12, 25].

Changes in chronic asthma are perpetuated because of a continuous dialogue between inflammatory cells and resident cells and matrix components in airways. The preferential recruitment of effector cells (eosinophils, Th2

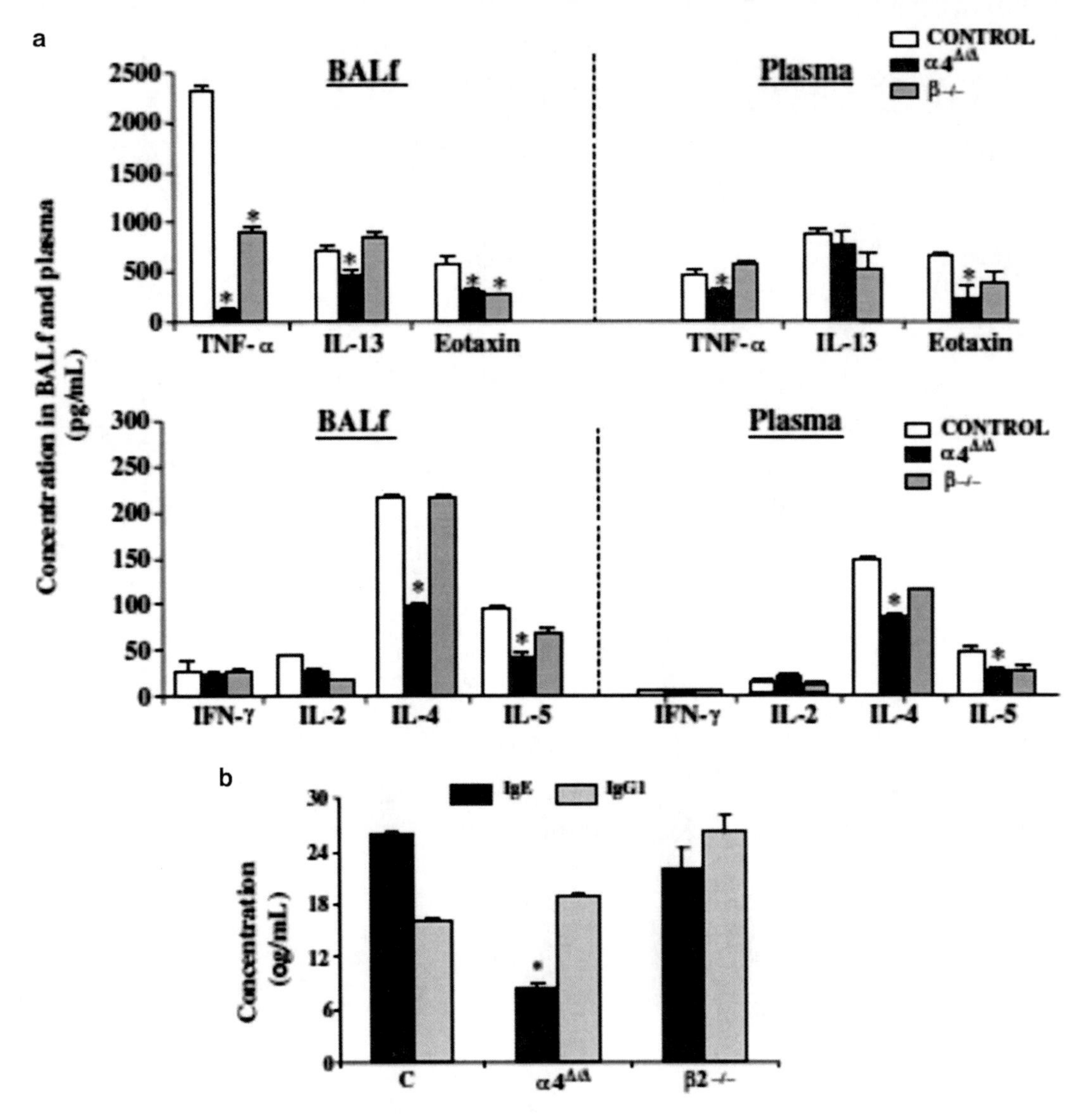

Fig. 4.4 (**a**) Concentration of cytokines (pg/mL) post-ovalbumin (OVA) (upper two panels) in bronchoalveolar lavage fluid (BALf) and plasma in control (*white bars*), $\alpha4^{\Delta/\Delta}$ (*black bars*), and $\beta2^{-/-}$ (*gray bars*) mice (*$p < 0.01$ compared to post-OVA control), and (**b**) concentrations of OVA-specific immunoglobulin (Ig) E and IgG1 (μg/mL) in plasma of OVA-treated mice in the three genotypes. Data (see Table 4.2) are averaged from two independent experiments, all assays were run in triplicate, $n = 8$ mice per group, and *$p < 0.01$ compared to post-OVA control values. *IFN-γ* interferon-γ,*IL* interleukin, *TNF-α* tumor necrosis factor-α

cells) in the lung and airways is mediated by a cascade of adhesive interactions initiated by activated endothelial cells. However, the distinct molecular pathways that dominate these processes are continuously being revised. Studies with anti-functional antibodies yielded conflicting data, depending on the type, dose, and route of administration of the antibody or on the animal model used [13–18, 20, 21, 26, 27]. Data with CD18-deficient mice suggested an absolute requirement of CD18 for recruitment of eosinophils in the airways and for AHR [19]. Our parallel comparison of CD18 with conditional α4Δ/Δ mice in acute asthma showed that, in the latter, not only was AHR and migration of cells to airway lumen

Table 4.2 Concentrations of cytokines in bronchoalveolar lavage fluid and plasma post-ovalbumin in control, $\alpha4^{\Delta/\Delta}$, and $\beta^{-/-}$ mice and concentrations of ovalbumin-specific immunoglobulins E and G_1 in plasma of ovalbumin-treated mice (data for Fig. 4.4)

	TNF-α (pg/mL)	IFN-γ (pg/mL)	IL-2 (pg/mL)	IL-4 (pg/mL)	IL-5 (pg/mL)	IL-13 (pg/mL)	Eotaxin (pg/mL).
BALf							
Control	2307.6 ± 34.98	26.7 ± 12.09	42.6 ± 3.97	216.0 ± 2.17	94.1 ± 2.87	713.8 ± 29.27	588.8 ± 65.44
$\alpha4^{\Delta/\Delta}$	**96.3 ± 2.8**	22.5 ± 1.08	24.9 ± 2.99	**96.4 ± 3.89**	**40.7 ± 7.09**	474.6 ± 90.59	**298.4 ± 16.8**
$\beta^{-/-}$	**909 ± 9.09**	26.4 ± 3.98	15.5 ± 1.5	214.2 ± 2.14	67.5 ± 6.75	855.8 ± 45.9	**274.2 ± 5.29**
Plasma							
Control	487.6 ± 8.47	3.5 ± 1.6	12.6 ± 1.98	149.2 ± 1.87	47.2 ± 4.87	875.5 ± 30.3	645.6 ± 23.97
$\alpha4^{\Delta/\Delta}$	**296.6 ± 2.8**	4.4 ± 1.8	18.3 ± 1.65	**843 ± 1.64**	**28.1 ± 2.75**	772.2 ± 118.3	**218.7 ± 122.06**
$\beta^{-/-}$	582.3 ± 5.3	3.3 ± 1.05	11.0 ± 0.64	113.6 ± 1.98	27.2 ± 4.56	535.1 ± 123	403.5 ± 74.90
IgE and IgG_1	IgE (μg/mL)	IgG_1 (μg/mL)					
Control	25.9 ± 0.25	15.9 ± 0.39					
$\alpha4^{\Delta/\Delta}$	**8.5 ± 0.43**	18.9 ± 0.42					
$\beta^{-/-}$	21.7 ± 2.65	26.1 ± 1.91					

Numbers in bold indicate $p < 0.01$

BALf bronchoalveolar lavage fluid, *IFN-γ* interferon-γ, *IgE* immunoglobulin E, *IgG_1* immunoglobulin G_1, *IL* interlukin, *TNF-α* tumor necrosis factor-α prevented, as in the CD18−/− mice, but there was attenuation of the sensitization process, minimal recruitment of eosinophils and lymphocytes in the lung parenchyma, and no upregulation in VCAM-1 expression [22]. As reported here, attenuation of all features of acute asthma observed in the absence of α4 integrins was, by and large, maintained during the chronic allergen challenge, although there was a higher leukocyte accumulation in the lungs of chronically challenged α4Δ/Δ mice (Figs. 4.1d and 4.2). Surprisingly, however, and in contrast to acute asthma, absence of β2 integrins (CD18) did not prevent chronic asthma development or its accompanied structural changes and AHR, despite the fact that migration of eosinophils to airways (BALf) (Figs. 4.1d and 4.2) was restrained in these mice. Although one cannot theoretically exclude the possibility that eosinophil degranulation differences between the two genetic models influence the cell numbers recorded, there is no experimental evidence supporting this notion. Thus, the migratory flow of eosinophils from the systemic circulation to lung and then through the lung interstitium to airway lumen may be dictated by differences in adhesive interactions between pulmonary vasculature and that of systemic bronchial circulation. Our data highlight the fact that α4 and β2 integrins play critical but nonredundant roles in facilitating this pathway of inflammatory cell migration from lung interstitium to airway lumen. Cooperativity between α4b1 (ligating domains 1 and 4 of VCAM-1) and aMβ2 (recognizing only domain 4 ofVCAM-1) in mediating adhesion and release from luminal ligands [23] may provide a mechanistic basis for such an outcome. However, the fact that the asthma phenotype was not prevented in VCAM-1-deficient mice [22] suggests that VCAM-1 upregulation and its interaction with integrins is not of critical importance in this process and that ligands other than VCAM-1 in airways and lung (i.e., E-selectin or fibronectin) may play pivotal roles in the absence of VCAM-1.

Total cell and total eosinophil accumulation in the lung was high in CD18−/− mice, even higher than in control mice. Such a high accumulation of inflammatory cells in the lung, not necessarily in BALf, may be contributing to chronic asthma development in these mice. However, our data with α4Δ/Δ mice would argue that mere accumulation of inflammatory cells, including eosinophils, in the lung might be necessary but not sufficient for asthma development. It is possible that the state of activation of accumulated cells is more important than their presence. For example, α4Δ/Δ mice did not show the elevated levels of TGF-β1, presumably because in the absence of α4 integrins, activation of the latent TGF-b1 in eosinophils (by mast-cell tryptase or avβ6 on epithelia [4]) was not induced [28]. Furthermore, in vitro studies have previously implicated a5b1 expressed by airway smooth muscle cells in regulating fibronectin deposition after TGF-b1 stimulation [29]. Thus, although the a5b1-dependent pathway is intact in α4Δ/Δ mice, the low levels of TGF-b1 are consistent with reduced fibrin deposition seen in α4Δ/Δ mice.

The role of eosinophils in acute or chronic asthma and their role in maintenance of chronic inflammation and airway remodeling is currently a subject of debate. Nevertheless, a large body of published evidence suggests a compelling role of eosinophils in chronic asthma, especially in humans [4]. Although caveats exist in the contribution of eosinophils in different strains of mice [30], our genetic models were of the same genetic background.

The recently appreciated role of anti-stem cell factor (SCF) antibody or oral imatinib mesylate in the attenuation of airway responses in both acute and chronic asthma is intriguing [31]. Interaction of fibroblasts with eosinophils that leads to increased production of cytokines is mediated through SCF [32]. Because of the known influence of SCF on integrin activation [33] and expression of SCF and c-kit in human asthmatic airways [34], it could be speculated that SCF effects are at least partially integrin dependent, as both c-kit and α4 integrin signalings are linked to the same pathways that regulate migration and activation of mast cells [35]. The role of mast cells, a constant component in allergic inflammatory response in the lung, has been controversial in airway disease development [36]. However, abundant experimental evidence in murine mod-

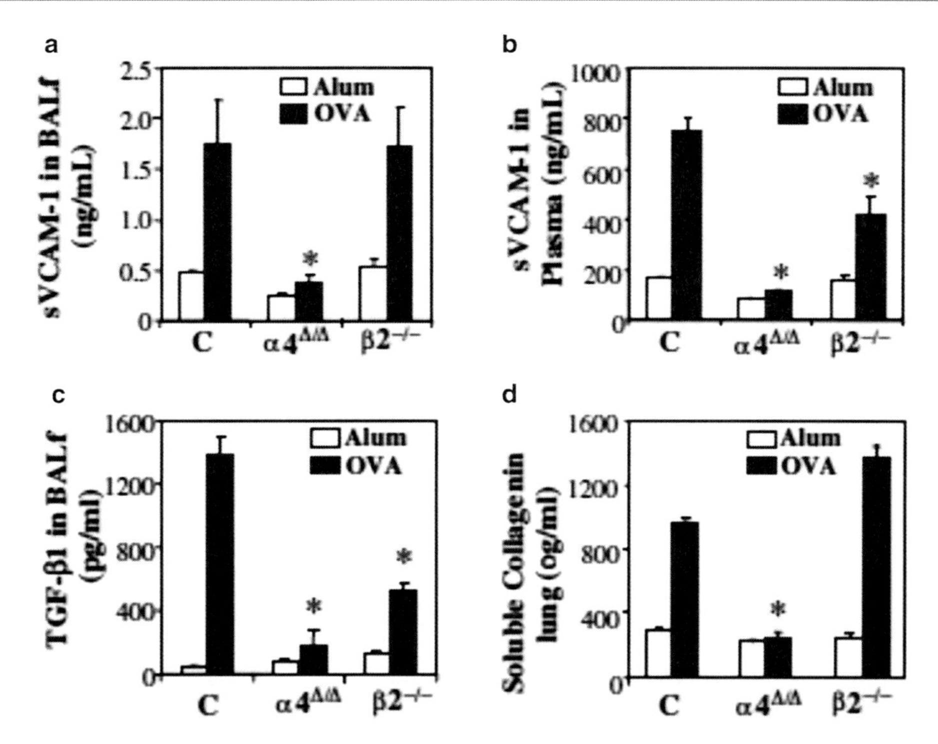

Fig. 4.5 (**a**) Soluble vascular cell adhesion molecule-1 (sVCAM-1) concentrations (ng/mL) in bronchoalveolar lavage fluid (BALf) and (**b**) in plasma measured by enzyme-linked immunosorbent assay (ELISA), (**c**) transforming growth factor-β1 (TGF- β1) levels (pg/mL) in BALf measured by ELISA, and (**d**) soluble collagen content (μg/mL) of lung measured by Sircol dye kit. Data are averaged from two independent experiments ± standard error of mean. All assays were run in triplicate, $n = 8$ mice/group, and $*p < 0.01$ compared to post-ovalbumin (OVA) control values

els has shown that mast cells adhere to mucosal surfaces through α4b1/α4b7 and VCAM-1 interactions [37] and secrete important mediators like TNF or CCL1, which can activate Th2 cells, or histamine and leukotriene B4 involved in recruitment of effector T cells in the lung [36].

Mice with mast-cell deficiency display marked reduction in lung mucosal inflammation, similar to mice with CCR8 deficiency or depletion of CD4 T lymphocytes [38], not unlike our findings with α4Δ/Δ mice. Also consistent with our findings, pretreatments with anti-very late antigen-4 antibody attenuated early response after OVA challenge through inhibition of mast-cell activation [39]. New insights about the mast-cell-dependent IL-17 activation and its role in asthma have been recently uncovered. IL-17 is thought to be produced by a distinct T-cell lineage (Th17 cells), and its production is negatively regulated by IFN-α and IL-4 [40, 41]. After antibody neutralization of IL-17 [42] or in IL-17 knockout mice [43], the OVA-induced initiation of asthma is prevented. Furthermore, mice not susceptible to asthma development (C3H) do not produce IL-17 [44]. These data suggest that IL-17 signaling is essential during antigen sensitization to establish asthma; however, in mice already sensitized, IL-17 seems to attenuate the allergic response [43]. It is of interest that, rather than Th17 cells, the main producers of IL-17 in asthma are alveolar macrophages [42]. Their activation and upregulation of IL-17A is mediated by products secreted by IgE/OVA-activated mast cells.

As the primary sensitization and OVA-specific IgE production in our α4−/− mice is significantly impaired, one may speculate that a decrease in IL-17A is expected in these mice. Although we have not measured levels of IL-17A in acute or chronic asthma in our mice, the ability of naïve

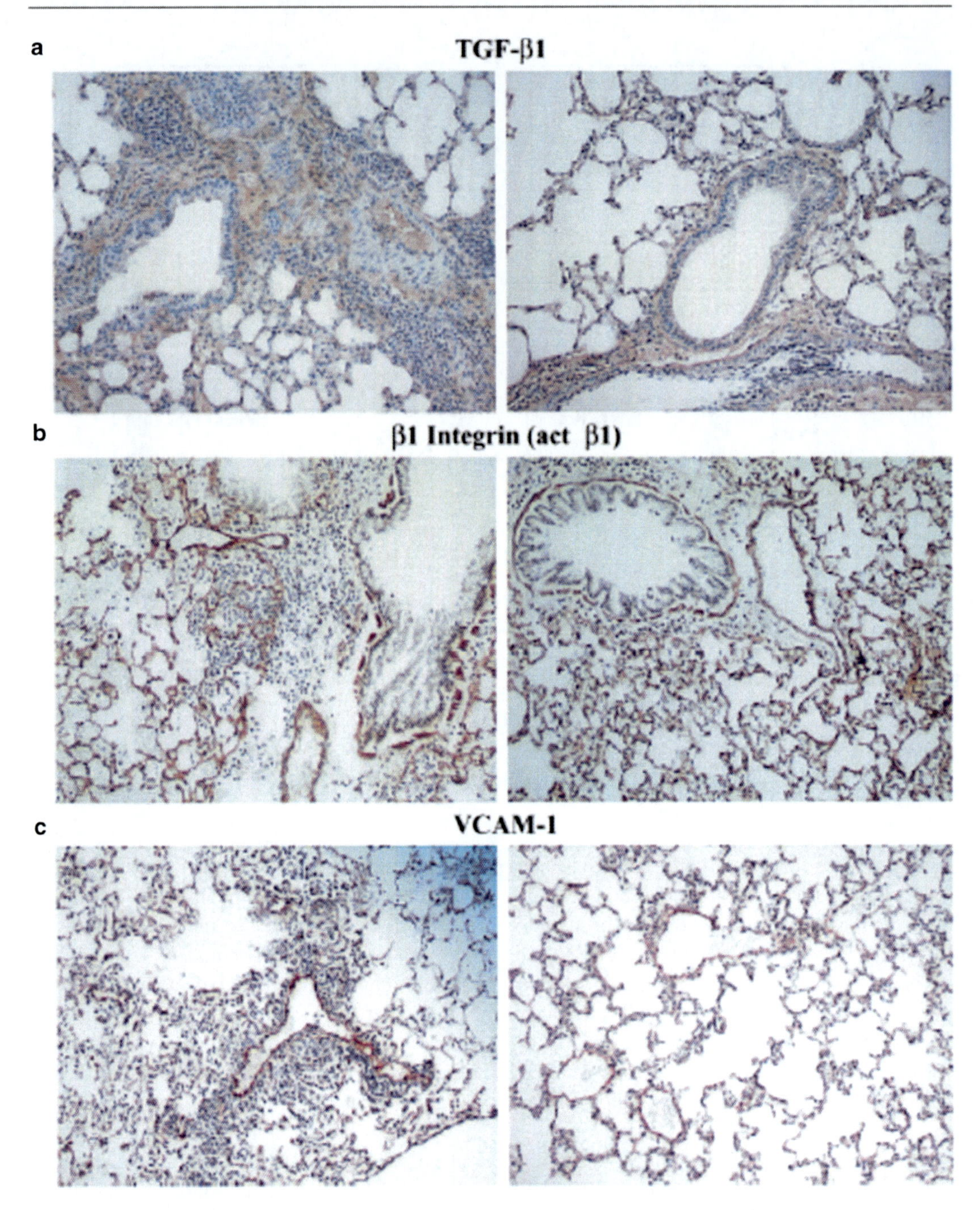

Fig. 4.6 Lung tissue sections, stained for transforming growth factor-β1 (TGF-β1) (anti-TGF-β1), 9EG7 (activated anti- β1 antibody), and vascular cell adhesion molecule-1 (VCAM-1) (anti-VCAM-1, MK/2) as described in Materials and Methods. (**a**) Images from control (*left*) and $\alpha4^{\Delta/\Delta}$ lung (*right*) post-ovalbumin (OVA) treatment and challenge stained with anti-TGF-β1. (**b**) Images of $\beta2^{-/-}$ lung (left) and $\alpha4^{\Delta/\Delta}$ lung (right) post-OVA stained with 9EG7. (**c**) Images of $\beta2^{-/-}$ lung (*left*) and $\alpha4^{\Delta/\Delta}$ lung (*right*) post-OVA stained with anti-VCAM-1. There were no significant differences in labeling with 9EG7, but labeling for TGF-β and VCAM-1 was less intense in $\alpha4^{\Delta/\Delta}$ mice

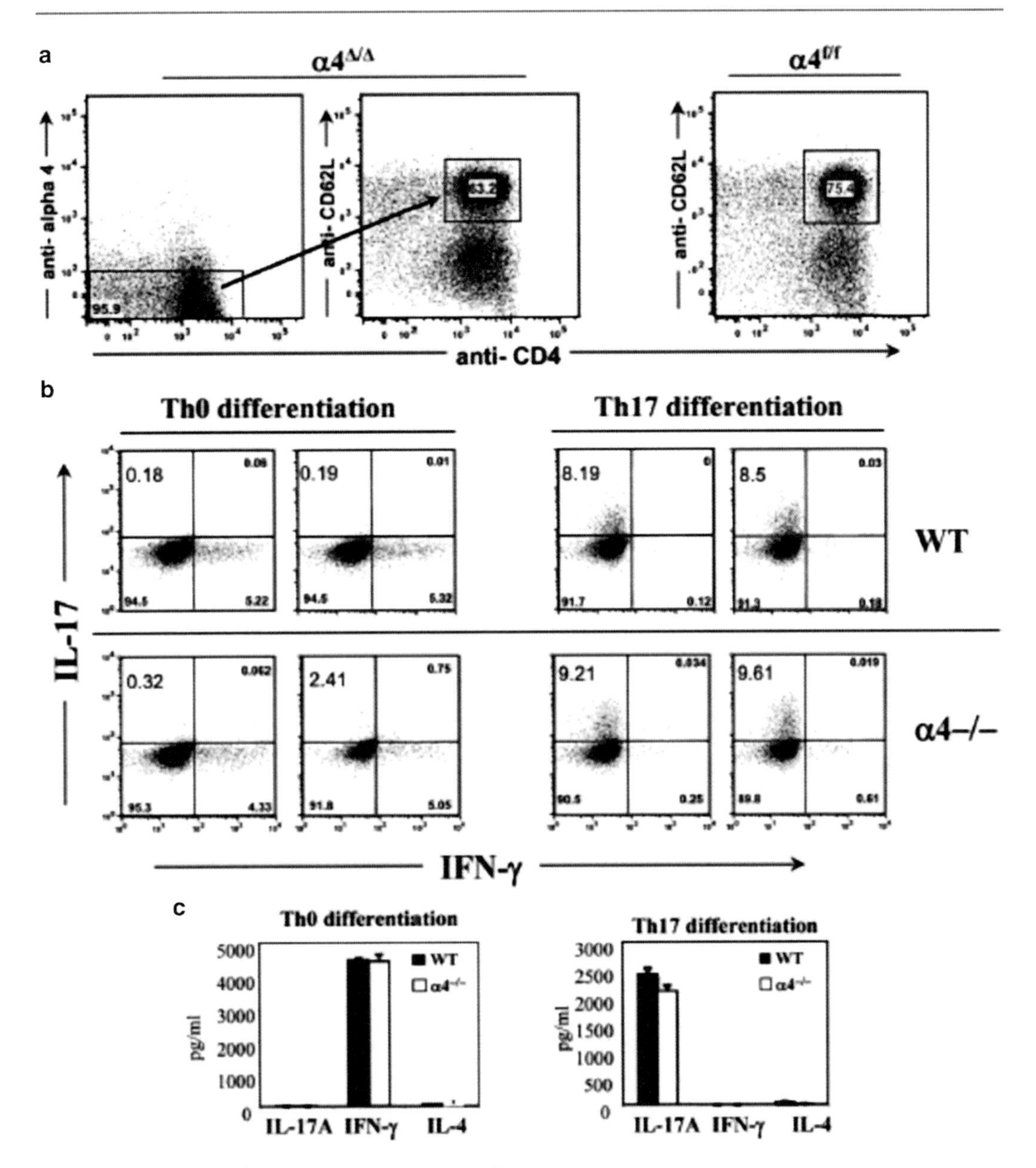

Fig. 4.7 Purified CD4$^+$/CD62L$^-$/α4$^-$_ and CD4$^+$/CD62L$^-$/α4$^+$ cells from spleens (**a**) were cultured in vitro under Th0 and Th17 differentiation conditions (see Materials and Methods). (**b**) Numbers in quadrants indicate the percentage of cells with intracellular interleukin (IL)-17 or interferon-γ (IFN-γ). There are no significant differences between WT and alpha 4 deficient in the proportion of cells expressing IL-17 or (**c**) in the levels of IL-17 in the supernatants analyzed for IFN-γ, IL-17A, and IL-4 by enzyme-linked immunosorbent assay

CD4+/CD62L+/α4- cells to produce IL-17 is not compromised (Fig. 4.7b, c) under in vitro conditions. Any influence of IL-17, if any, in our mouse model may be secondary to their impaired in vivo activation by mast cells or other cells and lack of migration of effector cells in BALf. Furthermore, induction of Foxp3+/CD103+ regulatory T cells in vitro under the influence of TGF-β is also not impaired in α4−/− T cells (18.3 % in α4+/+and 17.8 % in α4Δ/Δ).

Among cytokine levels in BALf, there is a striking reduction of TNF-α in α4Δ/Δ mice compared to other groups. Considerable in vitro and in vivo evidence suggests that TNF-α plays a key role in development of AHR, however, the molecular mechanisms linking TNF-α to AHR are not precisely defined [45]. Nevertheless, although we believe that the low levels found in α4Δ/Δ are likely contributing to lack of AHR in these mice, the reasons for the presence of low TNF-α are unclear and require further studies. IL-13 and its downstream target signal transducer and activator of transcription 6 exert a pivotal role in chronic asthma by affecting the function of resident airway cells (e.g., epithelia, fibroblasts, smooth muscle cells, mast cells) and, together with IL-4, by facilitating selective recruitment of inflammatory cells [46, 47].

Upregulation of VCAM-1 in airway vessels [14, 18] and increase of several chemokines (eotaxin or monocyte chemotactic protein-1, etc.) work in concert to establish chemotactic gradients between different compartments in the lung. In addition, direct participation of IL-13 in fibrosis has been advocated because of IL-13-mediated induction of arginase 1 or TGF-β and platelet-derived growth factor by endothelial cells and monocyte/macrophages and the effects of these mediators on fibroblast proliferation [48, 49]. Nevertheless, despite the critical role of IL-13 in chronic asthma [50], IL-13−/− mice still had AHR, albeit with less fibrosis, less inflammatory changes, and intraepithelial eosinophil accumulation, thus dissociating the presence of these changes from AHR [51]. Of interest, IL-13, along with other cytokines, was decreased in BALf of α4D/D mice, consistent with a decrease in total cells in BALf in these animals, but levels of IL-13 were not different in plasma compared to controls, highlighting the importance of local changes in cytokine/chemokine milieu for asthma development.

A phenotypic response similar to the one seen in our α4Δ/Δ animals in acute and chronic asthma was reported for sphingosine kinase inhibition [52] with suppression of eosinophil migration to airways, Th2 cytokine, and chemokine secretions and decreased AHR. Sphingosine 1-phosphate [52], like α4 integrin, is important in mast cell, neutrophil, and eosinophil chemotaxis, and its inhibition may have an effect on cellular migration machinery involving adhesion molecules.

It is of great interest that, despite the effectiveness of several inhibitors of integrins (i.e., antibodies, small molecules, peptides) in animal models of asthma (i.e., mouse, rat, sheep, guinea pig), human trials have been largely disappointing [23]. It is possible that movement of leukocytes or eosinophils in humans involves other molecules in addition to integrins, or that it involves a combination of integrins (i.e., both β1 and β2 integrins and possibly even some selectins) for dampening the inflammatory response. Furthermore, it should be pointed out that integrin inhibitors largely prevented disease in animals and may not be able to greatly influence already established disease in humans, although they should curtail either the frequency or severity of episodes. Along these lines, our data with the two genetic models (CD18−/−, α4Δ/Δ) may provide important preclinical relevance. It could be argued that a major impact of α4 integrin is to dampen the initial sensitization process. As a result, the effectiveness in established processes in humans may be less than that seen in animal models. Further, it is possible that even in the absence of additional inflammatory cell recruitment, the remodeling process persists through sustained activation of airway structural cells [53]. In addition, results we have reported with combined β2 and α4 deficiencies in leukocyte migration to inflammatory sites [54] are instructive in terms of the effectiveness of using combinations of integrin inhibition.

Overall, in chronic asthma there is parallel induction of a complex array of genes in a variety of cells recruited to the airways and in airway resident cells. The contribution of individual players (cells or cytokines) seems to be variable and under independent regulation [55]. We believe that our present studies addressing some of these issues by exploiting genetic murine models expand on the knowledge of the molecular understanding of the pathophysiology of chronic asthma.

Conclusion

α4 Integrins appear to have a regulatory role in directing transforming growth factor b-induced collagen deposition and structural alterations in lung architecture likely through interactions of Th2 cells, eosinophils, or mast cells with endothelium, resident airway cells, and/or extracellular matrix.

Materials and Methods

Animals

α4 integrin f/f mice were produced as described previously [24]. These mice were bred with Mxcre + mice, and the resulting Mxcre + α4 flox/flox mice were conditionally ablated as neonates by intraperitoneal injections of poly(I)poly(C) (Sigma Aldrich Co., St Louis, MO, USA) for interferon induction. cre_α4f/f mice were used as controls and the α4-ablated mice are referred to as α4Δ/Δ, and only mice with O 95 % α4 ablation in hematopoietic cells were used for studies. CD18 knockout mice were provided by Dr. Arthur Beaudet, Baylor College of Medicine (Houston, TX, USA). All animal protocols were approved by the University of Washington Institutional Animal Care and Use Committee. Mice were bred and maintained under specific pathogen-free conditions in University of Washington facilities and were provided with irradiated food and autoclaved water ad libitum.

Induction of Chronic Allergic Asthma

Mice were sensitized and later challenged with ovalbumin (OVA; Pierce Biotechnology, Inc., Rockford, IL, USA) as described previously [56]. Briefly, mice were immunized with 100 mg OVA complexed with aluminum sulfate in a 0.2-mL volume, administered by intraperitoneal injection on day 0. On day 8 (250 mg OVA) and on days 15, 18, and 21 (125 mg OVA), mice anesthetized briefly with inhalation of isoflurane in a standard anesthesia chamber were given OVA by intratracheal administration. Intratracheal challenges were done as described previously [57]. Mice were anesthetized and placed in a supine position on a board. The animal's tongue was extended with lined forceps, and 50 mL OVA (in the required concentration) was placed at the back of its tongue. We have previously shown that this protocol results in increased AHR, inflammation of the airways, and Th2 cytokine production [56–58]. Prolonged inflammation was induced by subsequent exposure of mice to 125 mg OVA intratracheally three times a week until groups of mice were sacrificed on day 55 (chronic phase) after the last intratracheal challenge on day 54 (Fig. 4.1a). The control group [α4+/+cre- mice also injected with poly(I) poly(C)] received normal saline with aluminum sulfate by intraperitoneal route on day 0 and 0.05 mL 0.9 % saline by intratracheal route on days 8, 15, 18, and 21 and three times a week until they were sacrificed on day 55.

Bronchoalveolar Lavage Fluid

Mice underwent exsanguination by intraorbital arterial bleeding and then lavaged (0.4 mL three times) from both lungs. Total bronchoalveolar lavage fluid (BALf) cells were counted from a 50-mL aliquot, and the remaining fluid was centrifuged at 200 g for 10 min at 4_C, and the supernatants were frozen for assay of BALf cytokines later. Cell pellets were resuspended in fetal bovine serum and smears were made on glass slides. The cells, after air-drying, were stained with Wright-Giemsa (Biochemical Sciences, Inc., Swedesboro, NJ, USA), and differential counts were enumerated using a light microscope at 40_ magnification. Cell number refers to that obtained from lavage of both lungs/mouse.

Lung Parenchyma Cell Recovery

Lung mincing and digestion was performed after lavage as described previously [17] with

100 mg/mL collagenase for 1 h at 37_C and filtered through a no. 60 sieve (Sigma Aldrich Co.). All numbers mentioned in this article refer to cells obtained from one lung/mouse.

Lung Histology

Lungs from other animals of the same group were fixed in 4 % paraformaldehyde overnight at 4_C. Tissues were embedded in paraffin and cut into 5-mm sections. A minimum of 15 fields were examined by light microscopy. The intensity of cellular infiltration around pulmonary blood vessels was assessed by hematoxylin and eosin staining. Airway mucus was identified by staining with Alcian blue and periodic acid Schiff staining as described previously [22]. Subepithelial pulmonary fibrosis was detected by Masson's trichrome and Martius scarlet blue stains as described in [59].

Lung Immunohistochemical Staining

Lungs were processed for immunohistochemical staining following standard procedure [57], and then stained with either anti-vascular cell adhesion molecule-1 (VCAM-1; MK/2), anti-b1 (9EG7), or antitransforming growth factor-b1 (TGF-b1). Briefly, tissues were fixed with 4 % paraformaldehyde in 100-mM phosphate-buffered saline (PBS; pH 7.4) for 6–12 h at 4°C; washed with PBS for 10 min three times and then soaked in 10 % sucrose in PBS for 2–3 h, 15 % sucrose in PBS for 2–3 h, 20 % for 3–12 h at 4°C; and then embedded in OCT compound (Tissue-Tek 4583; Sakura Finetechnical Co., Ltd, Tokyo, Japan) and frozen in acetone-cooled dry ice. Frozen blocks were cut on a freezing, sliding microtome at 4 mm (Leica CM1850 Cryostat) and air-dried for 30 min at room temperature (RT). After washing in PBS three times for 10 min at RT, 0.3 % hydrogen peroxide was applied to each section for 30 min at RT to block endogenous peroxidase activity. Each slide was incubated with blocking solution to block nonspecific reactions, and appropriately diluted primary antibody was applied to each slide and incubated overnight at 4°C. After washing with PBS, slides were incubated with appropriately diluted specific biotin-conjugated secondary antibody solution for 1 h at RT. After washing with PBS, slides were incubated in AB reagent for 1 h at RT (ABComplex/HRP; DAKO, Carpinteria, CA, USA), washed with PBS, and stained with 0.05 % 3,30-diaminobenzidine tetrahydrochloride (Sigma Aldrich Co.) in 0.05 M Tris buffer (pH 7.6) containing 0.01 % H_2O_2 for 5–40 min at RT. Slides were counterstained with Mayer's hematoxylin, dehydrated, and mounted.

Fluorescein-Activated Cell Sorter (FACS) Analysis

Cells from hemolyzed peripheral blood, bone marrow (BM), bronchoalveolar lavage, lung parenchyma, spleen, mesenteric lymph nodes, cervical lymph nodes, axillary lymph nodes, and inguinal lymph nodes were analyzed on a FACSCalibur (BD Immunocytometry Systems, San Jose, CA, USA) using the CellQuest program. Staining was performed by using antibodies conjugated to fluorescein isothiocyanate (FITC), phycoerythrin (PE), allophycocyanin (APC), peridinin-chlorophyll-protein (PerCP-Cy5.5), and Cy-chrome (PE-Cy5 and PE-Cy7). The following antibodies (BD Biosciences-Pharmingen, San Diego, CA, USA) were used for cell-surface staining: APC-conjugated CD45 (30 F-11), FITC-conjugated CD3(145-2C11), PE-Cy5-conjugated CD4 (RM4-5), PE-conjugated CD45RC (DNL-1.9), APC-conjugated CD8(53–6.7), PE-Cy5-conjugated β220 (RA3-6β2), FITC-conjugated IgM, PE-conjugated CD19 (ID3), PE-conjugated CD21(7G6), FITC-conjugated CD23 (B3B4), APC-conjugated GR-1 (RB6-8C5), and PE-conjugated Mac1(M1/70). PE-Cy5-conjugated F4/80 (Cl:A3-1[F4/80]) was obtained from Serotec Ltd. (Raleigh, NC, USA). PE-conjugated anti-α4 integrin (PS2) and anti-VCAM-1

(M/K-2) were from Southern Biotechnology (Birmingham, AL, USA). Irrelevant isotype-matched antibodies were used as controls. Among hematopoietic cells, CD45+/CD3+ were T cells, CD3+/CD4+ were helper T cells, and CD3+/CD8+ were cytotoxic T cells. B cells were B220+. Gr- 1+/F4/80_ cells were granular cells (e.g., neutrophils, eosinophils) and Gr-1-/ F4/80hi cells were tissue macrophages.

Cytokines

Cytokines (IL-2, IL-4, IL-5, tumor necrosis factor-a [TNF-α], and interferon-γ [IFN-γ]) in BALf and serum were assayed by FACS with Mouse Th1/Th2 Cytokine Cytometric Bead Assay (BD Biosciences) following manufacturer's protocol. Manufacturer's sensitivity for IL-2, IL-4, and IL-5 is 5 pg/mL, for IFN-γ is 2.5 pg/mL, and for TNF-α is 6.3 pg/mL. IL-13 and eotaxin were measured by enzyme-linked immunosorbent assay (ELISA) using Quantikine M kits (R&D Systems, Minneapolis, MN, USA), and the limit of detection is 1.5 pg/mL for IL-13 and 3 pg/mL for eotaxin.

OVA-Specific IgE and IgG1 in Plasma

Anti-mouse IgE (R35-72) and IgG1 (A85-1) from BD Biosciences were used for measuring OVA-specific IgE and IgG1, respectively, by standard ELISA procedures as described previously [60]. The lower and upper limits of detection for IgE and IgG1 are 3 ng/mL to 10 ug/mL (minimal detectable dose determined by adding 2 standard deviations of the mean OD 405 nm value for 20 replicates of the zero standard and calculating the corresponding concentration).

Pulmonary Fibrosis

Martius scarlet blue and Masson's trichrome stains in paraffin lung sections were used to visualize lung fibrosis [59].

Soluble VCAM-1 and Soluble Collagen in Lung Homogenate

Soluble VCAM-1 was determined as described previously [22]. Total amount of soluble collagen in the lung was measured using a Sircol collagen assay kit from Biocolor (Newtownsbury, Northern Ireland, UK), according to the method described [57]. In all experiments, a collagen standard was used to calibrate the assay.

Lung Function Testing

In vivo AHR to methacholine was measured 24 h after the last OVA challenge in conscious, free-moving, spontaneously breathing mice using whole-body plethysmography (model PLY 3211; Buxco Electronics, Sharon, CT, USA) as described previously [56]. Mice were challenged with aerosolized saline or increasing doses of methacholine (5, 20, and 40 mg/mL) generated by an ultrasonic nebulizer (DeVilbiss Health Care, Somerset, PA, USA) for 2 min. The degree of bronchoconstriction was expressed as enhanced pause, a calculated dimensionless value, which correlates with the measurement of airway resistance, impedance, and intrapleural pressure in the same mouse. Penh readings were taken and averaged for 4 min after each nebulization challenge. Penh was calculated as follows: Penh 5½ðTe = Tr_ 1+ _ ðPEF = PIF + _ where Te is expiration time, Tr is relaxation time, PEF is peak expiratory flow, and PIF is peak inspiratory flow _ 0.67 coefficient. The time for the box pressure to change from a maximum to a user-defined percentage of the maximum represents relaxation time. Tr measurement begins at the maximum box pressure and ends at 40 %.

Th2 Differentiation, Intracellular Staining, and ELISA Assay for IL-17A and IFN-γ

Cell suspensions from spleen were enriched for CD4+ T cells using anti-CD19 and anti-CD8 antibodies and BioMag goat anti-rat IgG

Fc beads (Qiagen Inc, Valencia, CA, USA). Naïve CD4+ CD62L + cells were sorted by flow cytometry, and α4+ cells were excluded from α4Δ/Δ cells (4.1 %). Five hundred thousand naïve CD4+ CD62L + T cells were cultured under Th0 conditions: anti-CD3 [1 mg/mL] and anti-CD28 [2 mg/mL]; Th17 polarization conditions, anti-CD3 [1 mg/mL], anti-CD28 [2 mg/mL], IL-6 [10 ng/mL], (Peprotech, Rocky Hill, NJ, USA), TGF-b [5 ng/mL], (Peprotech), anti-IL-4 [5 mg/mL] (clone 11B11), and anti-IFN-γ [5 mg/mL] (clone XMG1.2); or inducible regulatory T-cell conditions, anti-CD3 [1 mg/mL], anti-CD28 [2 mg/mL], TGF-b [5 ng/mL], anti-IL-4 [5 mg/mL], anti-IFN-γ [5 mg/mL], and IL-2 [100 U/mL] (eBioscience, Inc., San Diego, CA, USA). After 4 days, cells were restimulated with phorbol myristate acetate (PMA) (50 ng/mL; Sigma Aldrich) and ionomycin (1 mg/mL; Sigma Aldrich) and monensin (2 mM; eBioscience) for 5 h. Cells were stained with CD4-FITC and/or CD103-PE. Intracellular staining was performed using Perm and Fix solutions from eBioscience, anti-IL-17A-PE (clone eBio17B7), anti-IFN-γ-APC (clone XMG1.2), and anti-Foxp3-FITC. These antibodies were obtained from eBioscience. For ELISA analysis, supernatants were harvested after 4 days. IL-17A and IFN-γ levels were analyzed using monoclonal antibodies and recombinant cytokine standards from eBioscience. Detection limits were IL-17A (125 pg/mL) and IFN-γ (100 pg/mL).

Statistics

Mean 6 standard error of the mean was calculated using student's *t*-test in Excel Software (Microsoft, Redmond, WA, USA). A p value less than 0.01 was considered significant.

References

1. Wills-Karp M. Immunologic basis of antigen-induced airway hyperresponsiveness. Annu Rev Immunol. 1999;17:255–81.
2. Shum BO, Rolph MS, Sewell WA. Mechanisms in allergic airway inflammation-lessons from studies in the mouse. Expert Rev Mol Med. 2008;10:e15.
3. Elias JA, Zhu Z, Chupp G, Homer RJ. Airway remodeling in asthma. J Clin Invest. 1999;104:1001–6.
4. Kariyawasam HH, Robinson DS. The role of eosinophils in airway tissue remodelling in asthma. Curr Opin Immunol. 2007;19:681–6.
5. Benayoun L, Druilhe A, Dombret M-C, Aubier M, Pretolani M. Airway structural alterations selectively associated with severe asthma. Am J Respir Crit Care Med. 2003;167:1360–8.
6. Ebina M, Takahashi T, Chiba T, Motomiya M. Cellular hypertrophy and hyperplasia of airway smooth muscles underlying bronchial asthma: a 3-d morphometric study. Am Rev Respir Dis. 1993;148:720–6.
7. HoshinoM NY, Sim JJ. Expression of growth factors and remodeling of the airway wall in bronchial asthma. Thorax. 1998;53:21–7.
8. Jeffery PK. Remodeling in asthma and chronic obstructive lung disease. Am J Respir Crit Care Med. 2001;164:28S–38.
9. Payne DNR, Rogers AV, Adelroth E, et al. Early thickening of the reticular basement membrane in children with difficult asthma. Am J Respir Crit Care Med. 2003;167:78–82.
10. Tanaka H, Yamada G, Saikai T, et al. Increased airway vascularity in newly diagnosed asthma using a high-magnification bronchovideoscope. Am J Respir Crit Care Med. 2003;168:1495–9.
11. Zhou L, Li J, Goldsmith AM, et al. Human bronchial smooth muscle cell lines show a hypertrophic phenotype typical of severe asthma. Am J Respir Crit Care Med. 2004;169:703–11.
12. Epstein MM. Do mouse models of allergic asthma mimic clinical disease? Int Arch Allergy Immunol. 2004;133:84–100.
13. Borchers MT, Crosby J, Farmer S, et al. Blockade of CD49d inhibits allergic airway pathologies independent of effects on leukocyte recruitment. Am J Physiol Lung Cell Mol Physiol. 2001;280: L813–21.
14. Chin JE, Hatfield CA, Winterrowd GE, et al. Airway recruitment of leukocytes in mice is dependent on alphaα4-integrins and vascular cell adhesion molecule-1. Am J Physiol Lung Cell Mol Physiol. 1997;272:L219–29.
15. Henderson Jr WR, Chi EY, Albert RK, et al. Blockade of CD49d (alpha 4 integrin) on intrapulmonary but not circulating leukocytes inhibits airway inflammation and hyperresponsiveness in a mouse model of asthma. J Clin Invest. 1997;100:3083–92.
16. Koo GC, Shah K, Ding GJF, et al. A small molecule very late antigen- 4 antagonist can inhibit ovalbumin-induced lung inflammation. Am J Respir Crit Care Med. 2003;167:1400–9.
17. Laberge S, Rabb H, Issekutz T, Martin J. Role of VLA-4 and LFA-1 in allergen-induced airway hyperresponsiveness and lung inflammation in the rat. Am J Respir Crit Care Med. 1995;151:822–9.

18. Nakajima H, Sano H, Nishimura T, Yoshida S, Iwamoto I. Role of vascular cell adhesion molecule 1/very late activation antigen 4 and intercellular adhesion molecule 1/lymphocyte function-associated antigen 1 interactions in antigen-induced eosinophil and T cell recruitment into the tissue. J Exp Med. 1994;179:1145–54.
19. Lee S-H, Prince JE, Rais M, et al. Differential requirement for CD18 in T-helper effector homing. Nat Med. 2003;9:1281–6.
20. Kanwar S, Smith C, Shardonofsky F, Burns A. The role of MAC-1 (CD11b/CD18) in antigen-induced airway eosinophilia in mice. Am J Respir Cell Mol Biol. 2001;25:170–7.
21. Schneider T, Issekutz TB, Issekutz AC. The role of alpha 4 (CD49d) and beta 2 (CD18) integrins in eosinophil and neutrophil migration to allergic lung inflammation in the brown Norway rat. Am J Respir Cell Mol Biol. 1999;20:448–57.
22. Banerjee ER, Jiang Y, Henderson Jr WR, Scott LM, Papayannopoulou T. Alpha4 and beta2 integrins have nonredundant roles for asthma development, but for optimal allergen sensitization only alpha4 is critical. Exp Hematol. 2007;35:605–17.
23. Barthel SR, Johansson MW, McNamee DM, Mosher DF. Roles of integrin activation in eosinophil function and the eosinophilic inflammation of asthma. J Leukoc Biol. 2008;83:1–12.
24. Scott LM, Priestley GV, Papayannopoulou T. Deletion of alpha4 integrins from adult hematopoietic cells reveals roles in homeostasis, regeneration, and homing. Mol Cell Biol. 2003;23:9349–60.
25. Lloyd C, Gutierrez-Ramos J. Animal models to study chemokine receptor function: in vivo mouse models of allergic airway inflammation. Methods Mol Biol. 2004;239:199–210.
26. Larbi KY, Allen AR, Tam FWK, et al. VCAM-1 has a tissue-specific role in mediating interleukin-4-induced eosinophil accumulation in rat models: evidence for a dissociation between endothelial-cell VCAM-1 expression and a functional role in eosinophil migration. Blood. 2000;96:3601–9.
27. Lobb R, Hemler M. The pathophysiologic role of alpha 4 integrins in vivo. J Clin Invest. 1994;94:1722–8.
28. Banerjee ER, Latchman YE, Jiang Y, Priestley GV, Papayannopoulou T. Distinct changes in adult lymphopoiesis in Rag2(−/−) mice fully reconstituted by alpha4-deficient adult bone marrow cells. Exp Hematol. 2008;36:1004–13.
29. Moir LM, Burgess JK, Black JL. Transforming growth factor beta 1 increases fibronectin deposition through integrin receptor alpha 5 beta 1 on human airway smooth muscle. J Allergy Clin Immunol. 2008;121:1034–9. e1034.
30. Takeda K, Haczku A, Lee J, Irvin C, Gelfand E. Strain dependence of airway hyperresponsiveness reflects differences in eosinophil localization in the lung. Am J Physiol Lung Cell Mol Physiol. 2001;281:L394–402.
31. Berlin AA, Hogaboam CM, Lukacs NW. Inhibition of SCF attenuates peribronchial remodeling in chronic cockroach allergen-induced asthma. Lab Invest. 2006;86:557–65.
32. Dolgachev V, Berlin AA, Lukacs NW. Eosinophil activation of fibroblasts from chronic allergen-induced disease utilizes stem cell factor for phenotypic changes. Am J Pathol. 2008;172:68–76.
33. Kovach NL, Lin N, Yednock T, Harlan JM, Broudy VC. Stem cell factor modulates avidity of alpha 4 beta 1 and alpha 5 beta 1 integrins expressed on hematopoietic cell lines. Blood. 1995;85:159–67.
34. Al-Muhsen SZ, Shablovsky G, Olivenstein R, Mazer B, Hamid Q. The expression of stem cell factor and c-kit receptor in human asthmatic airways. Clin Exp Allergy. 2004;34:911–6.
35. Tan BL, Yazicioglu MN, Ingram D, et al. Genetic evidence for convergence of c-kit- and alpha 4 integrin-mediated signals on class IA PI-3kinase and the Rac pathway in regulating integrin-directed migration in mast cells. Blood. 2003;101:4725–32.
36. Reuter S, Taube C. Mast cells and the development of allergic airway disease. J Occup Med Toxicol. 2008;3 Suppl 1:S2.
37. Abonia JP, Hallgren J, Jones T, et al. Alpha-4 integrins and VCAM-1, but not MAdCAM-1, are essential for recruitment of mast cell progenitors to the inflamed lung. Blood. 2006;108:1588–94.
38. Gonzalo J-A, Qiu Y, Lora JM, et al. Coordinated involvement of mast cells and T cells in allergic mucosal inflammation: critical role of the CC chemokine ligand 1:CCR8 axis. J Immunol. 2007;179:1740–50.
39. Hojo M, Maghni K, Issekutz TB, Martin JG. Involvement of alpha −4 integrins in allergic airway responses and mast cell degranulation in vivo. Am J Respir Crit Care Med. 1998;158:1127–33.
40. Harrington LE, Hatton RD, Mangan PR, et al. Interleukin 17-producing CD4+ effector T cells develop via a lineage distinct from the T helper type 1 and 2 lineages. Nat Immunol. 2005;6:1123–32.
41. Park H, Li Z, Yang XO, et al. A distinct lineage of CD4 T cells regulates tissue inflammation by producing interleukin 17. Nat Immunol. 2005;6:1133–41.
42. Song C, Luo L, Lei Z, et al. IL-17-producing alveolar macrophages mediate allergic lung inflammation related to asthma. J Immunol. 2008;181:6117–24.
43. Schnyder-Candrian S, Togbe D, Couillin I, et al. Interleukin-17 is a negative regulator of established allergic asthma. J Exp Med. 2006;203:2715–25.
44. Lewkowich IP, Lajoie S, Clark JR, Herman NS, Sproles AA, Wills-Karp M. Allergen uptake, activation, and IL-23 production by pulmonary myeloid DCs drives airway hyperresponsiveness in asthma-susceptible mice. PLoS ONE. 2008;3:e3879.
45. Jain D, Keslacy S, Tliba O, et al. Essential role of IFNbeta and CD38 in TNFalpha-induced airway smooth muscle hyper-responsiveness. Immunobiology. 2008;213:499–509.

46. Kim BE, Leung DYM, Boguniewicz M, Howell MD. Loricrin and involucrin expression is down-regulated by Th2 cytokines through STAT-6. Clin Immunol. 2008;126:332.
47. Kumar RK, Herbert C, Webb DC, Li L, Foster PS. Effects of anticytokine therapy in a mouse model of chronic asthma. Am J Respir Crit Care Med. 2004;170:1043–8.
48. Broide DH. Immunologic and inflammatory mechanisms that drive asthma progression to remodeling. J Allergy Clin Immunol. 2008;121:560–70.
49. Lee CG, Homer RJ, Zhu Z, et al. Interleukin-13 induces tissue fibrosis by selectively stimulating and activating transforming growth factor beta (1). J Exp Med. 2001;194:809–22.
50. Nath P, Yee Leung S, Williams AS, et al. Complete inhibition of allergic airway inflammation and remodelling in quadruple IL-4/5/9/13 −/− mice. Clin Exp Allergy. 2007;37:1427–35.
51. Kumar RK, Herbert C, Yang M, Koskinen AML, McKenzie ANJ, Foster PS. Role of interleukin-13 in eosinophil accumulation and airway remodelling in a mouse model of chronic asthma. Clin Exp Allergy. 2002;32:1104–11.
52. Lai W-Q, Goh HH, Bao Z, Wong WSF, Melendez AJ, Leung BP. The role of sphingosine kinase in a murine model of allergic asthma. J Immunol. 2008;180:4323–9.
53. Roth M, Johnson PRA, Borger P, et al. Dysfunctional interaction of C/EBP{alpha} and the glucocorticoid receptor in asthmatic bronchial smooth-muscle cells. N Engl J Med. 2004;351:560–74.
54. Ulyanova T, Priestley G, Banerjee E, Papayannopoulou T. Unique and redundant roles of alphaα4 and beta2 integrins in kinetics of recruitment of lymphoid vs myeloid cell subsets to the inflamed peritoneum revealed by studies of genetically deficient mice. Exp Hematol. 2007;35:1256–65.
55. Koerner-Rettberg C, Doths S, Stroet A, Schwarze J. Reduced lung function in a chronic asthma model is associated with prolonged inflammation, but independent of peribronchial fibrosis. PLoS ONE. 2008;3:e1575.
56. Iwata A, Nishio K, Winn RK, Chi EY, Henderson Jr WR, Harlan JM. A broad-spectrum caspase inhibitor attenuates allergic airway inflammation in murine asthma model. J Immunol. 2003;170:3386–91.
57. Henderson Jr WR, Chi EY, Maliszewski CR. Soluble IL-4 receptor inhibits airway inflammation following allergen challenge in a mouse model of asthma. J Immunol. 2000;164:1086–95.
58. Henderson Jr WR, Lewis DB, Albert RK, et al. The importance of leukotrienes in airway inflammation in a mouse model of asthma. J Exp Med. 1996;184:1483–94.
59. Cho JY, Miller M, Baek KJ, et al. Inhibition of airway remodeling in IL-5-deficient mice. J Clin Invest. 2004;113:551–60.
60. Henderson Jr W, Banerjee E, Chi E. Differential effects of (s)- and (r)-enantiomers of albuterol in a mouse asthma model. J Allergy Clin Immunol. 2005;116:332–40.

Published in

Banerjee ER, Jiang Y, Henderson Jr WR, Latchman YL, Papayannopoulou T. Absence of $\alpha4$ but not $\beta2$ integrins restrains the development of chronic allergic asthma using mouse genetic models. Exp Hematol. 2009;37:715–27.

Role of Integrin α4 (VLA – Very Late Antigen 4) in Lymphopoiesis by Short- and Long-Term Transplantation Studies in Genetic Knockout Model of Mice

5

Abstract

Objective. α4 Integrins are major players in lymphoid cell trafficking and immune responses. However, their importance in lymphoid reconstitution and function, studied by antibody blockade or in genetic models of chimeric animals with α4KO embryonic stem (ES) cells, competitive repopulation experiments with fetal liver KO cells, or in b1/b7 doubly deficient mice has yielded disparate conclusions.

Materials and Methods. To study the role of α4 integrin (α4b1, α4b7) during adult life, we transplanted lethally irradiated Rag2−/− mice with α4Δ/Δ or α4f/f adult bone marrow (BM) cells and evaluated recipients at several points after transplantation.

Results. Lymphomyeloid repopulation (8 months later) was entirely donor-derived in all recipients, and novel insights regarding lymphoid reconstitution and function were revealed. Thymic repopulation was impaired in all α4Δ/Δ recipients, likely because of homing defects of BM-derived progenitors, although a role of α4 integrin in intrathymic expansion/maturation of T cells cannot be excluded; reconstitution of gut lymphoid tissue was also greatly diminished because of homing defects of α4Δ/Δ cells, impaired immunoglobulin (Ig) M and IgE, but normal IgG responses were seen, suggesting compromised initial B-/T-cell interactions, whereas interferon-g production from ovalbumin-stimulated cells was increased, possibly reflecting a bias against Th2 stimulation.

Conclusion. These data complement previous observations by defending the role of α4 integrin in thymic and gut lymphoid tissue homing and by strengthening evidence of attenuated B-cell responses in α4-deficient mice.

E.R. Banerjee, *Perspectives in Inflammation Biology*,
DOI 10.1007/978-81-322-1578-3_5,

Introduction

Integrins are heterodimeric (a/b) cell-adhesion receptors with important roles in many physiologic or pathological cell processes, including cell migration and lymphocyte trafficking during homeostasis, recruitment of leukocytes in inflammatory sites, or metastatic spread of leukemic or tumor cells [1, 2]. Their effects are achieved not only through adhesion-dependent processes but also bidirectional signaling (outside-in, inside-out) and cross talk with other cell receptors (receptor tyrosine kinases or chemokine receptors) or signaling molecules, contributing to the generation of a wide diversity of signals [3]. Among integrins, α4b1 (very late antigen 4 [VLA4], CD49d/CD29), expressed in both hematopoietic and nonhematopoietic cells, and β2 (CD18) integrins, expressed only in hematopoietic cells, are major players in cell-migration events and lymphocyte trafficking. Circulating leukocytes are recruited to inflammatory tissues through interactions with tissue-selective adhesion/migration cascades. α4b1 participates in all the classic three steps of the trafficking cascade (rolling/adhesion/migration), in contrast to many molecules participating only at specific steps in this cascade, and it can further augment cell migration by other integrins in a transdominant fashion [4, 5].

Lymphocyte migration to secondary lymphoid tissues is necessary for maintaining immune defense and is regulated by multiple adhesion cascades controlled by shared participation of β1 and β2 integrins. Stromal cell networks in lymphoid tissues serve as guides directing or limiting the migration of T and B cells in and out of these tissues. Recently, there have been significant advances in defining the trafficking signals that control the movement of distinct subsets of immune cells in and out of specific tissue sites [6, 7]. Thymus function and maintenance of its population relies on the continuous supply of bone marrow (BM)-derived lymphoid progenitors. Significant knowledge in dissecting the molecular cascades that dictate the thymic tropism of BM-derived progenitors, the characterization of progenitors endowed with thymic tropism, and the intrathymic trafficking molecules responsible for intrathymic differentiation has recently been gained. Common lymphoid progenitors (CLP; Lin–/Sca1lo/IL-7Ra+) can give rise to B, T, natural killer (NK), and dendritic cells, but these are not present in thymus and need to be differentiated further within BM before their emigration to thymus. In fact, it is the CLP-2 population (c-kit–/B220+/CD19–) that possesses thymic tropism, and the latter is supported by both the lymphocyte function-associated antigen-1 (CD11a/CD18) and α4 integrin with the participation of GTP-binding protein coupled-receptor engagement [7]. Within the thymus, in addition to lymphocyte function-associated antigen-1 and α4b1, a5b1, CD44, and the CCL9/CCR25 chemokine pathway are working in concert for migration of developing thymocytes [7, 8], whereas other molecules (i.e., S1P/S1PR) control egress from thymus [9]. Thus, maintenance of thymic population in the adult is dependent on the coordination of several events, i.e., generation of BM-derived progenitors endowed with thymic tropism, their proper emigration and homing to thymus, and their appropriate intrathymic development. Defects along any of these steps significantly impair thymic population.

Genetic studies with α4 null chimeric mice have uncovered an important role of α4b1 integrin in lymphopoiesis and myelopoiesis with a severe defect in B-cell differentiation [10]. In addition, emigration of T-cell precursors from BM and their homing to thymus was inhibited in these mice, corroborating a large body of antibody blockade studies against α4 integrins or vascular cell-adhesion molecule (VCAM)-1 [11]. However, recent competitive transplantation experiments using fetal α4 null or α4+/+ cells with adult BM a+/+ competitor cells concluded that thymic repopulation was normal, but impairment in Peyer's patches (PPs) reconstitution was seen like in the chimeras [12]. Further, in contrast to the evidence from α4 null chimeric mice, b1 integrin-induced deletion in adult hematopoietic cells proved not to be essential for hematopoiesis/lymphopoiesis and for lymphocyte trafficking with normal homing to lymph nodes and PPs [13]. Only a transient defect in thymic colonization and/or intrathymic differentiation of T cells was seen in these mice, but an impairment in immunoglobulin M (IgM) response was present [13]. Thus, there are great discrepancies in data obtained in different genetic models of α4 integrin deficiency, or in different experimental settings used, and data with antibody blockade have not been predictive of results obtained in models of genetic deficiencies.

We have previously described adult mice with conditional ablation of α4 integrin and studied its effects on hematopoiesis [14]. No gross hematopoietic defects were seen in our model, but a sustained alteration in biodistribution of progenitor and stem cells at homeostasis was seen and α4-ablated cells had a competitive disadvantage in long-term hematopoietic repopulation [15]. However, no detailed studies on lymphopoiesis and lymphoid cell function were previously done in these mice. In the present studies, we transplanted α4-ablated BM cells from adult mice into lethally irradiated Rag2 null mice and made detailed observations in lymphoid organ repopulation and lymphocyte function in fully donor-reconstituted recipients. We have uncovered distinct defects in thymus reconstitution, in homing to gut lymphoid tissue, and in IgM-mediated responses.

Similarities and differences with previously used models are discussed, with an attempt to reconcile divergence in outcomes and further our understanding of the role of α4 integrin in adult lymphopoiesis.

Materials and Methods

Mice

Generation of α4f/f and MxCre + α4f/f mice [14] or Tie2Cre + α4Δ/Δ mice was previously described [16]. Rag 2–/– (CD45.1) mice were from Taconic (Germantown, NY, USA). For transplantation experiments using these mice, single-cell suspensions of 5×10^6 donor BM cells in prewarmed Hanks balanced salt solution were injected via tail veins into each of the lethally irradiated (800 cGy) Rag2–/– recipients (n=10/group). Recipient animals were studied 8–10 weeks and up to 8 months posttransplantation.

Antibodies and Fluorescein-Activated Cell-Sorting (FACS) Evaluation

Nucleated cells from donor or recipient mice were analyzed using CellQuest software on a FACSCalibur (BD Immunocytometry Systems, San Jose, CA, USA). Antibodies and their clone numbers included CD3 (145-2C11), CD8 (53–6.7), CD25 (7D4), CD44 (KM114), CD28 (37.51), B220 (RA3-6B2), Gr-1 (RB6-8C5), and α4b7 (DATK32) purchased from BD Biosciences (San Diego, CA, USA); anti-α4 integrin (PS/2 from Southern Biotech, Birmingham, AL, USA), CD19 (ID3) from Serotec Ltd. (Raleigh, NC, USA), CD34 (RAM34) from Caltag/Invitrogen (Carlsbad, CA, USA); and CD62L (MEL-14) in addition to mouse regulatory T-cell staining kit (w/ PE Foxp3 FJK-16 s, fluorescein isothiocyanate CD4, allophycocyanin CD25) from eBioscience, San Diego, CA, USA. Irrelevant isotype-matched antibodies (BD Biosciences) were used as controls.

Preparation of Tissues for Cellularity and FACS Evaluation

BM, spleen, thymus, PPs, mesenteric lymph nodes (MLNs), cervical lymph nodes, axillary lymph nodes, and inguinal lymph nodes were used for studies. The lymph nodes (LN) from the anatomical locations mentioned here were surgically removed, and the sacs were gently teased using two bent syringe needles in Dulbecco's phosphate-buffered saline (PBS) with 0.1 % bovine serum albumin (BSA) on ice. Thymus glands were gently rubbed between the rough surfaces of two histological slides until only a cell suspension remained. Single-cell suspensions were made by gently pushing through narrow gauge needles once or twice, and then debris or large-membrane particles from the sac were removed by passing through 40-mm Nitex filter (Sefar America, Depew, NY, USA), centrifuged, then resuspended in fresh PBS plus 0.1 % BSA. For PP cellularity assessment, the junction between the small and large intestines (segment of w2.5 cm) was surgically removed, and gut contents were removed, washed repeatedly, cut into small pieces, and digested with 0.1 % collagenase type I (Sigma Chemical Co., St. Louis, MO, USA) for 1 h in a 37_C water bath with periodic vortexing. The resulting cell suspension was washed in PBS to remove the enzyme, filtered through a nylon mesh, and then resuspended in PBS plus 0.1 % BSA for staining and evaluation by FACS. The subset distribution among CD45+ cells was determined. Cellularity was determined using a Particle Z cell counter from Beckman Coulter (Miami, FL, USA).

Immunohistochemistry

Tissues for immunohistochemistry were processed as described previously [17].

Immunization with Trinitrophenyl Ovalbumin (TNP-OVA)

TNP-OVA (Biosearch Technologies, Novato, CA, USA) and incomplete Freud's adjuvant (Sigma Chemical Co.) were emulsified and injected (1:1 ratio) 100/mg in 100/mL subcutaneously in the back of each mouse. Mice were bled 7 and 14 days later, and TNP-antibody titers in serum were measured by enzyme-linked immunosorbent assay (Sigma Chemical Co.). Goat anti-mouse IgM-alkaline phosphatase (AP), IgG1-AP, IgG2a-AP, and IgG3-AP, and rat anti-mouse IgE-AP were used for detection of various antibody isotypes. Calculated endpoint titers represent the greatest dilution of plasma with a signal (optical density) of 10 % of maximum [18].

Proliferative Responses and Cytokine Secretions by Lymphoid Cells

Splenocytes were prepared from spleens and cell suspensions were treated with red cell lysis buffer (Tris-NH4Cl). CD4+ or CD8+ T cells were purified from splenocytes by positive selection magnetic-activated cell sorting (Miltenyi Biotec, Auburn, CA, USA) (95 % purity) and stimulated with anti-CD28 (4 mg/mL) in the presence of various concentrations of anti-CD3 for 3 days. Antigen-presenting cells (APCs) were purified by Thy1.1 (Miltenyi Biotec) depletion using magnetic-activated cell-sorting columns and were activated with various concentrations of lipopolysaccharide (Sigma) or anti-IgM (BD Biosciences). All cell cultures were performed in RPMI-1640 with proper supplements. To assay proliferation, cultures were pulsed with 1 mCi/well of tritiated thymidine (Perkin Elmer, formerly New England Nuclear, Waltham, MA, USA) for the last 6 h of the 72-h incubation period. To measure cytokines, aliquots of supernatants were harvested at 72 h after initiation of cultures. Interleukin-4 (IL-4) and interferon-γ (IFN-γ) were analyzed by enzyme-linked immunosorbent assay using monoclonal antibodies and recombinant cytokine standards from eBioscience. Detection units were interleukin (IL)-4, 40 pg/mL, and IFN-γ, 100 pg/mL. Antigen-specific responses were detected using OVA-specific transgenic T-cell (OT-11) mice. CD4+ OT-11 T cells and APCs were isolated as above. The 1×10^5 OT-11 T cells were cultured with 3×10^5 g-irradiated (3000 cGy) APCs and various concentrations of OVA323-339 peptide (Invitrogen). Proliferation and cytokine production was measured as mentioned previously.

Results

Hemopoietic Reconstitution by Donor Cells in Rag 2−/− Recipients

Because transplantation of nonirradiated Rag 2−/− mice leads to very low hematopoietic

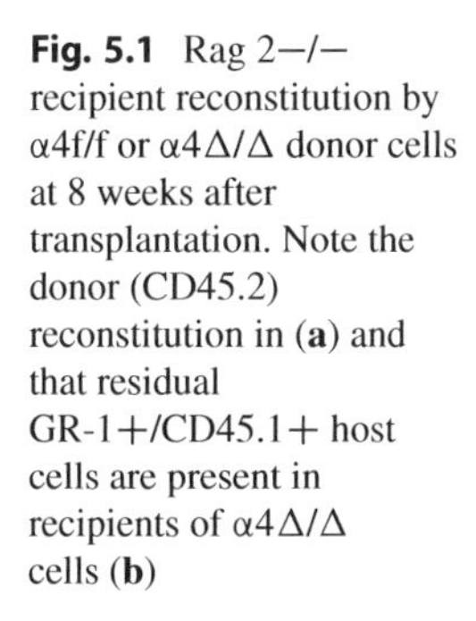

Fig. 5.1 Rag 2−/− recipient reconstitution by α4f/f or α4Δ/Δ donor cells at 8 weeks after transplantation. Note the donor (CD45.2) reconstitution in (**a**) and that residual GR-1+/CD45.1+ host cells are present in recipients of α4Δ/Δ cells (**b**)

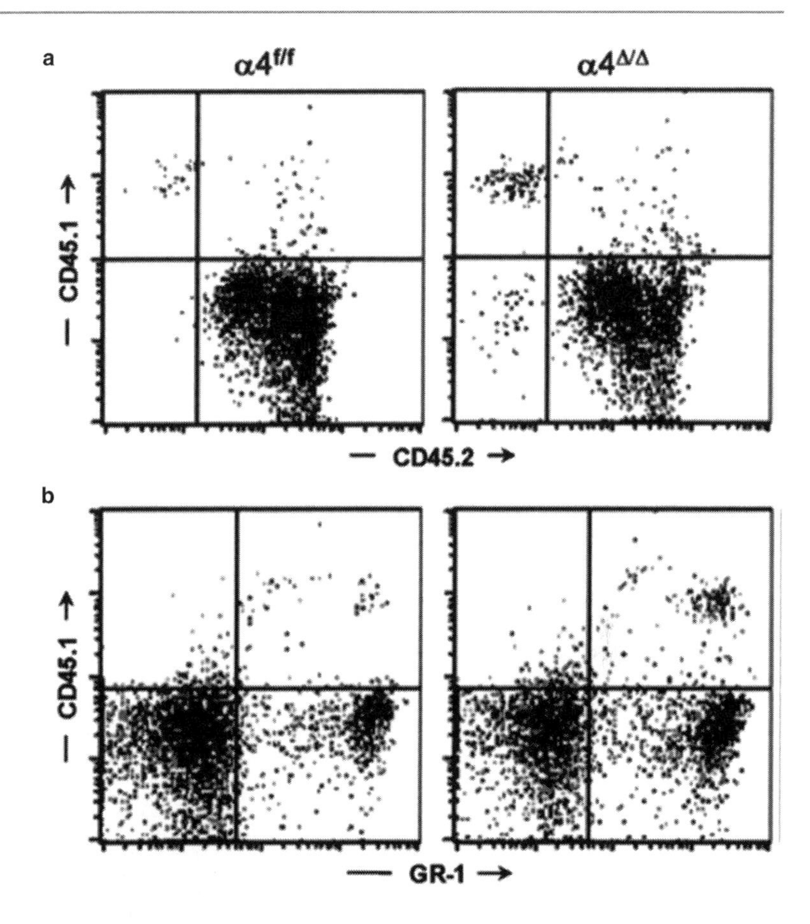

Table 5.1 Peripheral blood 8 weeks after transplantation

Donor cell type	$CD3^+$ (% $\alpha4^+$)	B220 (% $\alpha4^+$)	Gr-1 (% $\alpha4^+$)
$\alpha4^{f/f}$	28.3 ± 3.4 (99.7)	45.5 ± 32.5 (96.3)	27.2 ± 3.1 (97.8)
$\alpha4^{\Delta/\Delta}$	33.2 ± 4.8 (1.3)	45.0 ± 2.8 (1.6)	18.5 ± 2.5 (2.11)

Values are percentages. Lymphoid and myeloid reconstitution at 8 weeks after transplantation

reconstitution and to a largely ineffective thymic repopulation by donor cells [19], we used lethally irradiated Rag 2−/− recipients in two independent experiments using 20 mice given either control (α4f/f, 10 mice) or α4-deficient (α4Δ/Δ, 10 mice) donor BM cells. Expression of α4 in the donor population was O95 % in the controls and < 3 % in the α4Δ/Δ BM cells. A total of 5×10^6 donor cells were transplanted by tail-vein injection within 3–4 h after irradiation in the recipient mice. To assess reconstitution by donor cells, we tested recipient mice from 8 to 34 weeks posttransplantation. Of the 10 recipients given α4Δ/Δ cells, 1 died at 5 days and 1 at 119 days posttransplantation (with BM hypoplasia). When mice were tested at 8–10 weeks posttransplantation, the cohort of Rag 2−/− recipients given α4+/+ cells was completely reconstituted by donor cells (CD45.2+ cells), whereas among recipients of α4Δ/Δ cells there were O7 % residual host cells α4+ (CD45.1+) compared to 2 % in control recipients (Fig. 5.1 and Table 5.1). These data of somewhat delayed reconstitution by α4-deficient cells are consistent with homing and reconstitution defects seen previously in transplantation experiments with α4-deficient donor cells [14]. At 6 months, half of the recipient

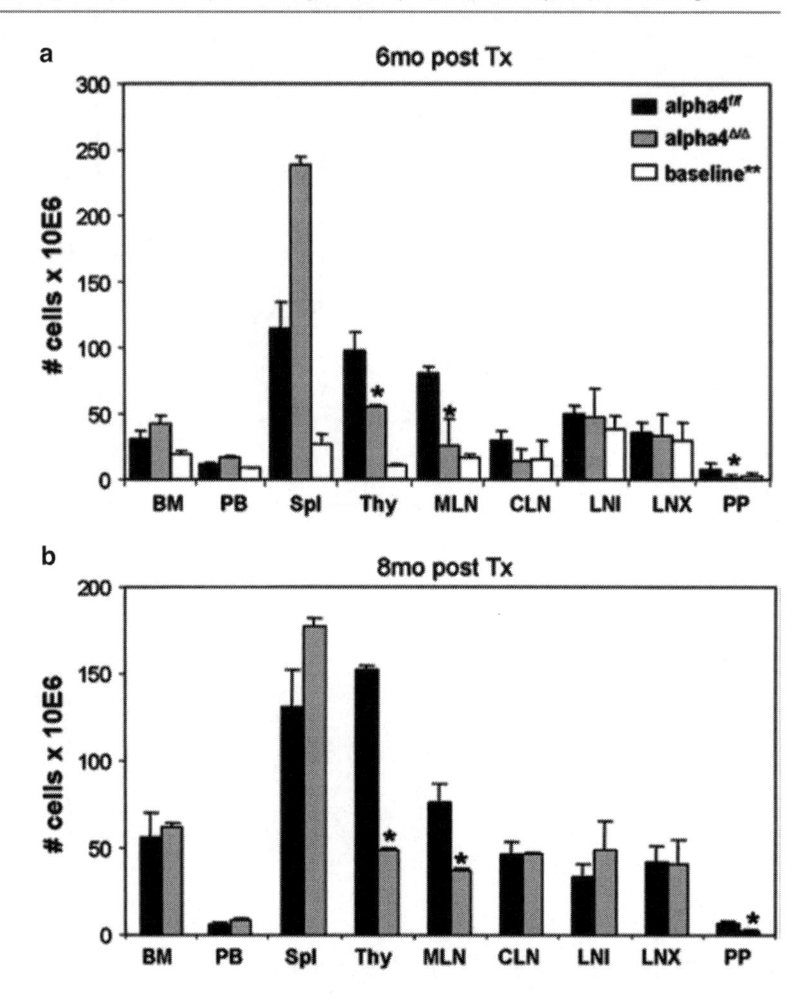

Fig. 5.2 Cell numbers recovered from different tissues at 6 and 8 months posttransplantation. *BM* bone marrow, *CLN* cervical lymph nodes, *LNI* inguinal lymph nodes, *LNX* axillary lymph nodes, *MLN* mesenteric lymph nodes, *PB* peripheral blood, *PP* Peyer's patches, *Thy* thymus. Baseline refers to nucleated cell counts in nontransplanted Rag2−/− mice. $*p < 0.05$

mice of either donor cells were sacrificed to study cellularity and differentiation parameters in all hemopoietic organs. Data from BM, peripheral blood (PB), and spleen are presented in Fig. 5.2a. In all these tissues, there was similar to increased cellularity in the recipients of α4Δ/Δ cells compared to the control group. Comparable data were obtained when the rest of recipient mice were sacrificed at 8 months posttransplantation, suggesting that this pattern is stably maintained long term (Fig. 5.2b). The complete donor cell reconstitution for each cohort was confirmed not only by replacement posttransplantation by CD45.2 (donor) cells but also by the level of α4 positivity in Gr-1+ cells in recipients of α4f/f versus α4Δ/Δ donor cells (BM α4+, 86.8 % vs. 0.88 %; PB, 89 % vs. 3.3 %; spleen, 78.8 % vs. 15.35 % cells, respectively). Presence of differentiated erythro-myeloid cells (Gr-1+, TER119+, Mac-1) in all these tissues was not significantly different between the two groups; however, differences were seen in the proportion of lymphoid cells. For example, there were decreased proportions of B220+ cells in the BM of α4Δ/Δ compared to α4f/f repopulated mice (19.5 % vs. 25.0 %, respectively). Among B-cell subsets, pro B (B220+/CD34–) and especially mature B cells (β220+/CD19+ or IgM+) were significantly decreased in BM of α4Δ/Δ recipients (Table 5.2). In PB the opposite was seen, with all B220+ cells being increased, especially the early types (B220+/CD19– or B220+/CD34+) in the recipients of α4Δ/Δ cells. Total T cells in BM

Table 5.2 Total cellularity (× 10E6) in hematopoietic tissues of $Rag2^{-/-}$ recipients (pooled data from 6 to 8 months posttransplantation)

	Donor cells	CD3+	CD4+	CD8+	CD4+ CD25+	CD3+ CD**44**+	Ratio CD**4**:CD8	B220+ CD**34**+	CD**34**–	CD19–	CD19+	Ig**M**+
BM	$\alpha 4^{f/f}$	9.05 ± 1.**45**	6.03 ± 1.0**54**	3.02 ± 0.08	**4**.57 ± 1.18	7.7 ± 2.09	1.99:1	0.7 ± 0.15	2.07 ± 0.8**4**	1.9**4** ± 0.05	3.3 ± 1.07	0.58 ± 0.02
	$\alpha 4\Delta/\Delta$	13.2 ± 3.05	9.32 ± 1.0**54**	3.**84** ± 0.65	**1.37** ± 1.07	**3.9** ± 1.13	2.**42**:1	0.9 ± 0.17	**0.15** ± 0.04	0.**29** ± 0.08	**1.56** ± 0.**4**3	0.67 ± 0.01
PB	$\alpha 4^{f/f}$	3.96 ± 0.68	2.**64** ± 0.08	1.32 ± 0.08	1.59 ± 0.78	2.9 ± 0.13	2.**4**:1	0.03 ± 0.001	0.39 ± 0.67	0.05 ± 0.001	0.34 ± 0.11	2.05 ± 0.22
	$\alpha 4\Delta/\Delta$	5.2 ± 0.11	2.6 ± 0.31	2.52 ± 0.06	1.7 ± 0.5	**4.69** ± 1.05	1.08:1	**0.14** ± 0.07	**1.34** ± 0.01	**0.46** ± 0.001	**1.34** ± 0.62	1.77 ± 0.03
Spleen	$\alpha 4^{f/f}$	13.8 ± 3.97	9.8 ± 1.97	3.9 ± 1.06	6.2 ± 0.17	12.3 ± 2.06	2.5:1	10.91 ± 3.38	0.62 ± 0.0**4**	1.09 ± 0.06	10.**4** ± 3.31	28.8 ± **4**.82
	$\alpha 4\Delta/\Delta$	**36.5** ± 6.14	**28.4** ± 3.14	**8.1** ± 1.033	**13.5** ± 3.91	**35.1** ± **4.47**	3.5:1	**39.23** ± 1.19	**3.83** ± 1.83	**4.96** ± 0.67	**71.2** ± 14.21	**51.62** ± 2.33
PP	$\alpha 4^{f/f}$	1.04 ± 0.09	0.8 ± 0.01	0.17 ± 0.01	0.85 ± 0.27	0.82 ± 0.25	**4**.7:1	11.07 ± 3.91	0.096 ± 0.003	0.005 ± 0.001	0.8 ± 0.01	1.32 ± 0.**4**3
	$\alpha 4\Delta/\Delta$	0.06 ± 0.02	0.5 ± 0.02	0.1 ± 0.003	0.29 ± 0.11	**0.56** ± 0.13	5:1	2.98 ± 0.93	**3.35** ± 1.39	**0.03** ± 0.01	**0.2** ± 0.01	0.826 ± 0.13
MLN	$\alpha 4^{f/f}$	28.59 ± 3.06	19.96 ± 3.07	9.53 ± 1.08	18.0 ± 2.**47**	**24.4** ± 2.21	2.09:1	7.87 ± 1.09	0.71 ± 0.21	0.12 ± 0.0**4**	**4**.86 ± 2.91	21.8**4** ± 3.96
	$\alpha 4\Delta/\Delta$	**8.7** ± 1.02	6.8 ± 1.02	1.86 ± 0.07	**2.9** ± 1.07	**6.15** ± 1.93	3.**4**:1	6.59 ± 1.17	**0.5** ± 0.03	**0.08** ± 0.002	**1.5** ± 0.3	**10.92** ± 2.49
PLN	$\alpha 4^{f/f}$	9.8 ± 1.07	5.1 ± 0.07	**4**.6 ± 0.**74**	3.9 ± 0.03	5.1 ± 1.**84**	1.1:1	1.8 ± 0.75	0.71 ± 0.05	0.65 ± 0.02	3.0 ± 0.75	6.1 ± 2.92
	$\alpha 4\Delta/\Delta$	9.**4** ± 1.95	6.8 ± 0.95	2.5 ± 0.0**4**	**1.66** ± 0.75	3.1 ± 1.16	2.72:1	1.**4**7 ± 0.28	1.06 ± 0.03	0.61 ± 0.03	2.7 ± 0.35	7.55 ± 1.12

BM bone marrow, *MLN* mesenteric lymph nodes, *PB* peripheral blood, *PLN* peripheral lymph nodes, *PP* Peyer's Patches
Numbers in bold letters: $p < 0.05$

were increased, but the proportion of activated T cells (CD4+ CD25+ regulatory cells, CD3+ CD44+–activated T cells) was diminished (the former from 50.5 % in α4f/f vs. 10.4 % in α4Δ/Δ and the latter from 85.6 % in α4f/f to 20.5 % in α4Δ/Δ). In addition to cell numbers, we tested progenitor cells colony-forming unit-culture (CFU-C) in all these tissues in the same recipients. Like the increased cellularity, a significant increase in progenitor content was found, especially in spleen and in peripheral blood (Fig. 5.3). The increase in progenitors concerned all subtypes (burst-forming unit erythroid, CFU-granulocyte-macrophage, and CFU-granulocyte-erythrocyte-megakaryocyte-monocyte, data not shown). It is of interest that the above quantitative changes in Rag 2–/– recipients of cells were similar to those we previously described for donor α4Δ/Δ mice [14], including increased numbers of B cells in circulation, although mature B cells (IgM+) were much less represented. These data do suggest that for this phenotype exemplified by changes in progenitor biodistribution and early release of B cells in circulation, the absence of α4 only in hematopoietic cells is necessary and sufficient [16].

Repopulation of Lymphoid Organs with α4Δ/Δ Donor Cells

Having documented that the Rag 2–/– recipients were fully reconstituted by donor (CD45.2) cells, we next assessed the repopulation status of several lymphoid tissues. We tested thymus, peripheral lymph nodes (inguinal, axillary, and cervical), and gut lymphoid tissues, i.e., PPs and MLNs.

Thymus

Repopulation of thymus was significantly impaired in Rag 2–/– recipients of α4Δ/Δ cells, compared to those that received α4f/f donor cells (Fig. 5.2a, b). A decrease in total cellularity (by 43 % at 6 months) was again demonstrable at 8 months (Fig. 5.2b) posttransplantation, indicating no restorative evidence with time posttransplantation. To test whether the decrease in cellularity concerned only certain subsets of lymphoid cells versus all cell types, we carried out FACS analyses using several lymphoid-specific markers with differential expression at different stages of activation or differentiation. These data are presented in Fig. 5.4 and Table 5.2. Double-positive (DP, CD4+/CD8+) population was the predominant one in Rag2–/– recipients of α4Δ/Δ or α4f/f donor cells. The CD4:CD8 ratio greatly favored the CD4 population (w8:1). Thus, the data in Figs. 5.2 and 5.4 and Table 5.2 suggest that the total repopulation of thymus was impaired, likely because of impaired migration of BM-derived progenitors to thymus, although their subsequent maturation (to DP) was not grossly impaired in the absence of α4 integrins. However, it is notable that CD8+ cells were at very low levels in thymus and lower than controls, in contrast to levels in PB (w1.9:1, Table 5.3).

Peripheral Lymph Nodes

Cellularity in cervical, axillary, and inguinal lymph nodes was similar to controls (i.e., recipients of α4f/f donor cells). Detailed evaluation of subset distribution showed that there were modestly decreased proportions of mature B cells (B220+IgM+) in all LNs tested, or decreased proportions of activated T cells (CD3+/CD25+, CD3+/CD44+), but their absolute numbers were not significantly different from control groups (Table 5.2). There was a tendency for preferential migration of CD45RC–/CD4+ (memory) cells to lymph nodes, whereas CD45RC+/CD4+ (naïve) cells instead preferentially migrated to spleen and thymus in α4Δ/Δ recipients. In spleen, as noted above, the cellularity, especially of red pulp, was significantly increased (Fig. 5.2) and concerned all developmental stages of B cells and of total T cells (Table 5.2). The picture of T- and B-cell distribution in spleen is more in line with what is present in PB and contrasts that of BM (Table 5.2) described above most likely because of longer retention and maturation of these cells

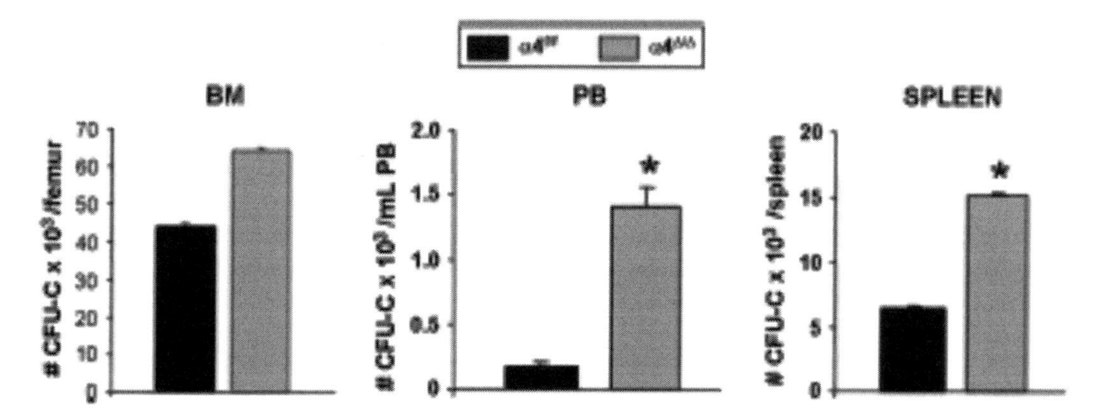

Fig. 5.3 Number of progenitor cells in bone marrow (*BM*), peripheral blood (*PB*), and spleen in Rag $2^{-/-}$ recipients of $\alpha4^{f/f}$ or $\alpha4^{\Delta/\Delta}$ cells at 6 months posttransplantation. Burst-forming erythroid, colony-forming unit (*CFU*) granulocyte-macrophage, CFU-Mix present are pooled and shown as CFU-C. No significant differences in proportions of different types of progenitors are seen. $*p < 0.05$

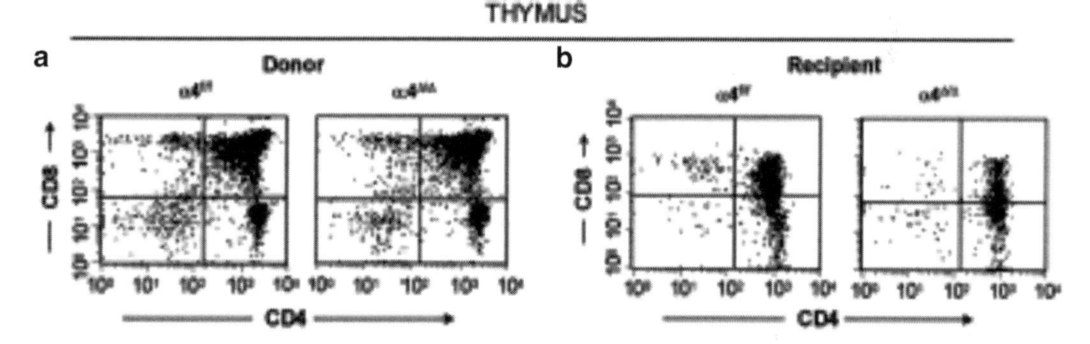

Fig. 5.4 Fluorescein-activated cell-sorting analyses of cells from thymi of donor ($\alpha4^{\Delta/\Delta}$, $\alpha4^{f/f}$) compared to recipient mice (of $\alpha4^{\Delta/\Delta}$ cells, $\alpha4^{f/f}$ cells) at 6 months posttransplantation. Note the low abundance of CD4$^+$, CD8$^+$, or double-negative populations and high numbers of double-positive cells in all recipient (Rag $2^{-/-}$) mice, indicating suboptimal reconstitution of thymus in transplanted Rag $2^{-/-}$ mice

Table 5.3 Thymus

	%DN	%DP	%CD4 SP	%CD8 SP	Ratio CD4:CD8
α4f/f ($n = 6$)	7.06 ± 2.72	48.67 ± 5.37	39.28 ± 5.05	5.0 ± 0.51	7.8:1
α4Δ/Δ ($n = 3$)	5.89 ± 2.06	45.61 ± 4.10	43.46 ± 1.55	5.04 ± 1.20	8.7:1

in the splenic environment compared to BM. CD40+ (dendritic cells) were lower in all organs except the spleen, where the proportion, but not the total number, was low (data not shown).

PPs and MLNs

In Rag−/− recipients of α4Δ/Δ cells both at 6 and 8 months posttransplantation, there was a significant reduction in cell numbers recovered from these tissues compared to controls (about 17-fold in PPs, 67 % less in MLNs) (Fig. 5.2). All subsets of B and T cells (Table 5.2) were severely reduced in PPs and MLNs repopulated by α4Δ/Δ cells. The CD4:CD8 ratio in PPs favored a CD4+ profile, as seen in thymus. These data, like the ones in thymus, suggest significant homing impairment of all α4Δ/Δ cells (Table 5.2) to these tissues.

Functional Status of α4-Deficient Lymphoid Cells

B Cells. To test the ability of α4-deficient B cells for antibody production, recipient mice were injected with TNP-OVA (intraperitoneally), and antibody titers were tested 7 and 14 days later. Determination of TNP-OVA-specific endpoint titers showed a significant reduction of titers for IgM (endpoint titers 1/2269.167 vs. 1/858.199 in α4Δ/Δ) and IgE (titers: 1/8248.8 vs. 1/1187.127 in α4Δ/Δ) in α4Δ/Δ recipients' sera, but no significant differences in IgG1 and IgG3 from α4f/f recipient's sera (Fig. 5.5a). This impairment in IgM response suggests that α4-integrin-dependent signals are necessary for the initial B-cell activation and IgM secretion. Because IgG1 and IgG3 or total IgG levels were not impaired, the data suggest no global defect in immune response or in Ig class switch. Further in vitro testing of B-cell proliferation supports this notion. Indeed, there was no defect in the in vitro activation of α4-deficient APCs (CD19+/CD3–) in the presence of increasing concentrations of lipopolysaccharide or anti-IgM (data not shown). In addition, there was no defect in activation and cytokine production by OT-II T cells (α4+) when OVA peptide was presented by α4Δ/Δ APCs as compared to α4f/f APCs (data not shown).

T Cells

The proliferative response of purified CD4+ or CD8+ or total splenocytes from α4Δ/Δ recipients was assessed by measuring proliferation in the presence of anti-CD3 and anti-CD28, or irradiated syngeneic APCs (for costimulation). Using both modalities, we identified no differences in proliferation of purified CD4+, CD8+, or total splenocytes between the α4-deficient and control test population. Also, no differences were seen after phorbol myristate acetate or Ionomycin stimulation (data not shown). To test whether secretion of cytokines by stimulated cells was normal, we measured under both costimulatory conditions (i.e., CD28 or irradiated APCs) the secretion of IFN-γ, IL- 2, IL-4, and IL-10. We found significantly decreased levels of IFN-γ by freshly isolated α4-deficient CD4+ T cells (Fig. 5.5b, left panel). There were low to undetectable levels of IL-2, IL-4, and IL-10 in these cultures stimulated with anti-CD3 and anti-CD28. To investigate whether the reduced IFN-γ production was due to an intrinsic defect of α4Δ/Δ CD4+ T cells, CD4 + CD25– (naïve) cells were sorted from α4Δ/Δ spleens and activated with anti-CD3 and anti-CD28 under Th1-polarizing (+IFN-γ) and Th2-polarizing (+IL-4) conditions. The percentage of α4Δ/Δ naïve CD4+ T cells producing IFNγ under Th1-polarizing conditions was similar to control CD4+ T cells (Fig. 5.5c). In addition, the percentage of IL-4-producing cells under Th2 conditions was comparable. Similar results were seen when CD4 + CD62L + or negative cells were sorted and activated with anti-CD3 and anti-CD28 (data not shown). This indicates there is no inherent defect by naïve α4Δ/Δ CD4 cells to produce IFN-γ. To test whether α4Δ/Δ can also mount a good response after in vivo stimulation, we immunized mice with OVA (7 days previously) and restimulated with anti-CD3 + anti-CD28 or OVA. As seen in Fig. 5.5b, in contrast to the response of naïve α4-deficient T cells, the OVA-activated α4-deficient T cells produced higher than control levels of IFN-γ in two independent experiments when restimulated with anti-CD3 + anti-CD28 (Fig. 5.5b, middle panel). Similar results were also seen when splenocytes from mice immunized with OVA were restimulated with OVA (Fig. 5.5b, right panel).

These data may indicate that more antigen-specific cells were maintained in the spleens of α4Δ/Δ mice, as the proliferative response to OVA was also increased (data not shown and [20]). However, further studies are needed to secure this point. Although not tested in Rag 2–/– recipients, we demonstrated no differences in the proportion of T regulatory (Tregs, CD4 + CD25+, FoxP3+) cells in donor α4-ablated mice,both at young (4 months old) and old (16 months) age (Figs. 5.6 and 5.7).

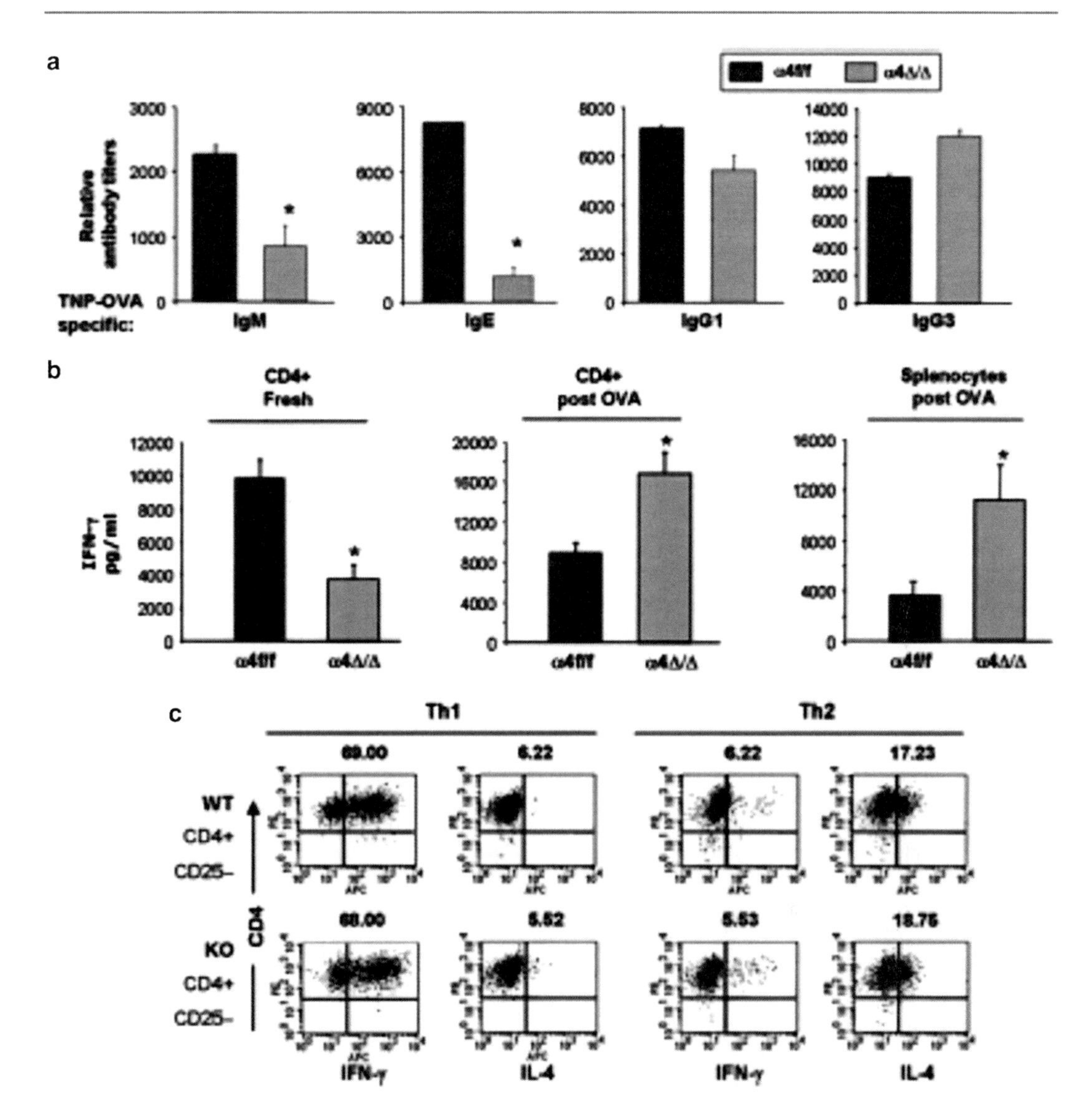

Fig. 5.5 Functional responses of B and T cells in recipients of $\alpha4^{f/f}$ or $\alpha4\Delta/\Delta$ donor cells. (**a**) Trinitrophenyl ovalbumin (TNP-OVA)–specific immunoglobulins in response to in vivo stimulation by TNP-OVA in recipients of $\alpha4^{f/f}$ or $\alpha4^{\Delta/\Delta}$ donor cells. (**b**) *Left panel*: interferon-γ (IFN-γ) production by freshly isolated $CD4^+$ T cells from recipients of $\alpha4^{f/f}$ or $\alpha4^{\Delta/\Delta}$ donor cells stimulated with anti-CD3 and anti-CD28. Middle panel: $\alpha4^{f/f}$ or $\alpha4^{\Delta/\Delta}$ mice were immunized with OVA and 7 days later, $CD4^+$ T cells were isolated from spleen, restimulated with anti-CD3 and anti-CD28, then IFN-γ production measured. *Right panel*: mice were stimulated with OVA and 7 days later, splenocytes were restimulated with OVA, then IFN-γ production measured. (**c**) $CD4^+CD25^-$ naïve cells were sorted and tested under Th1 conditions: anti-CD3/anti-CD28/anti-IL-4/IFN-γ or under Th2 conditions: anti-CD3/anti-CD28/anti-IFN-γ/IL-4, cultured for 7 days with IL-2 and restimulated with phorbol myristate acetate/ionomycin. Cytokines were detected by intracellular staining. Numbers above the upper right quadrant indicate the percentage of $CD4^+$ T cells producing each cytokine. $*p < 0.05$

Discussion

In the present studies, in contrast to previous observations using anti-VLA4 monoclonal antibodies [21] or transplant experiments of mixed chimeras with embryonic stem (ES) cells or fetal cells [10, 12, 22], we assessed lymphomyeloid hematopoietic reconstitution in lethally irradiated Rag 2−/− recipients transplanted with adult BM α4-deficient (α4Δ/Δ) donor cells. As the

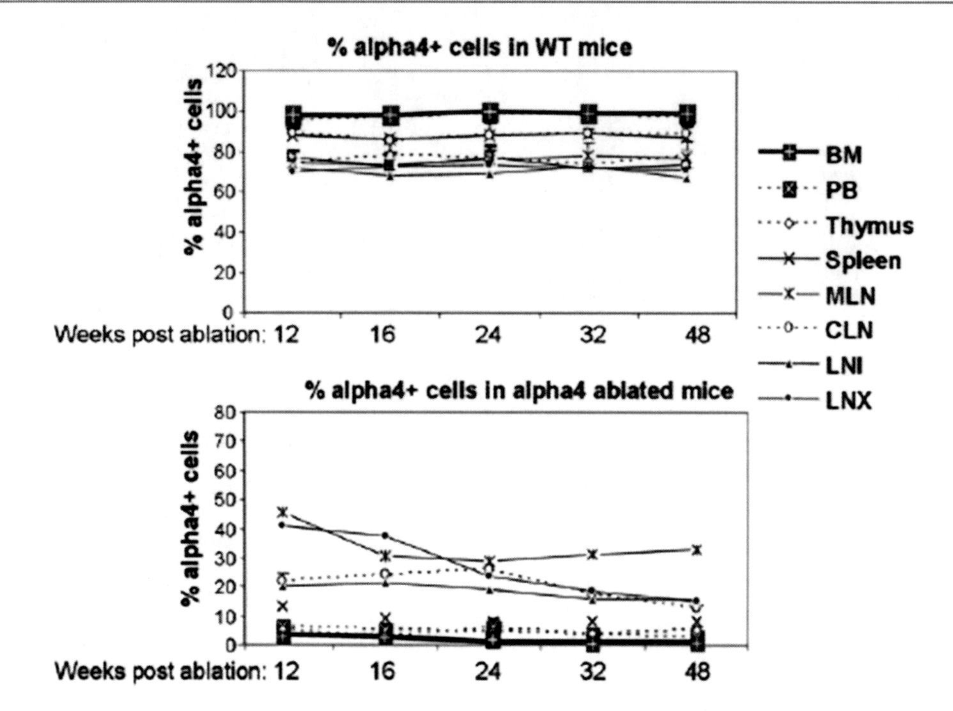

Fig. 5.6 Distribution of $\alpha 4^{+}$ cells in primary and secondary lymphoid organs from donor Mx*cre*α4Δ/Δ versus control mice. Single-cell suspensions from various organs were isolated, stained with anti-α4 antibody, and then analyzed by fluorescein-activated cell sorting, $n = 5$/group. Note the incomplete replacement by α4-negative cells (hematopoietic cells + stroma) in some lymphoid tissues (i.e., thymus, mesenteric lymph nodes), in contrast to bone marrow and peripheral blood. *BM* bone marrow, *CLN* cervical lymph nodes, *LNI* inguinal lymph nodes, *LNX* axillary lymph nodes, *MLN* mesenteric lymph nodes, *PB* peripheral blood

phenotype of transplanted animals may be different in the presence of normal host cells [22] or with normal competitor cells [12] because of paracrine or other undefined effects of normal companion cells, meaningful comparisons to the phenotype of donor animals were not feasible in previous studies. The complete reconstitution of hematopoiesis by α4-deficient BM donor in our Rag2−/− recipients had similar general features to that achieved in other recipients of adult α4-deficient cells [14]. These data reinforce our previous conclusions that the cellular composition and the premature, ongoing release of progenitors from BM to blood are dictated by the absence of α4 integrin in hematopoietic cells, with no demonstrable contribution by α4-deficient microenvironmental cells [16]. Reconstitution of lymphoid organs in the recipients of α4Δ/Δ cells and comparison to data in α4-ablated donor mice several months postablation were not studied previously. Such an evaluation in the present study revealed a constellation of novel, insightful findings with both similarities and differences from other relevant models, i.e., chimeras with α4KO ES cells, or competitive repopulation experiments using fetal liver α4KO cells [12, 22].

Repopulation of adult thymus was thought to be dependent on ongoing colonization by BM-derived progenitors [6, 7]. Consistent with this view was the finding of atrophic thymi postnatally in the chimeras with α4KO cells [22]. Because general hematopoiesis was not contributed by α4KO ES cells beyond the first month of postnatal life in these chimeric studies [23], the thymic homing competency of the BM-derived progenitors could not be tested in this model. Subsequent transplantation studies of fetal liver α4KO cells suggested no significant defects in thymic repopulation of the recipient mice [12].

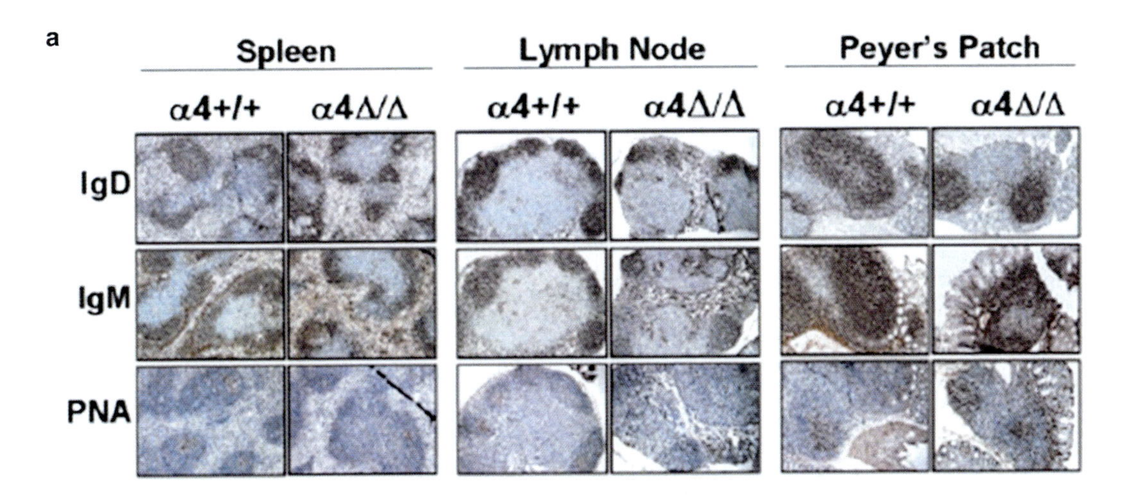

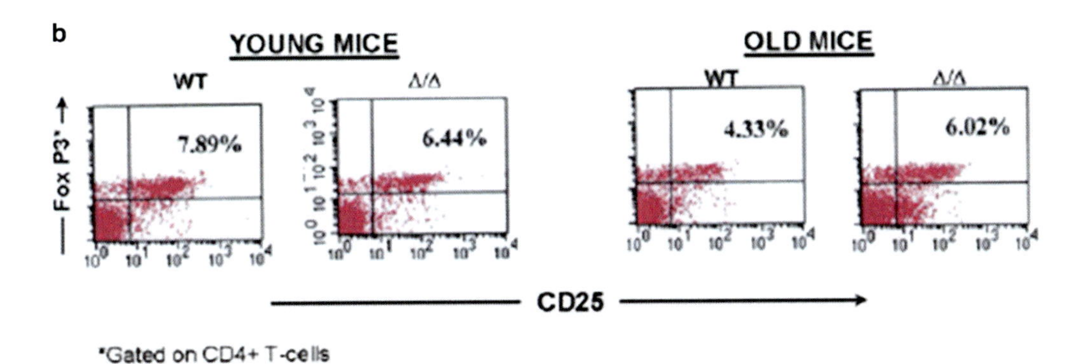

Fig. 5.7 (**a**) Immunohistochemistry staining of spleens, lymph nodes, and Peyer's patches from $\alpha4^{+/+}$ and $^{\Delta/\Delta}$ mice with anti-immunoglobulin (Ig)D, anti-IgM, and peanut agglutinin. No gross differences were noted except the presence of smaller lymphoid nodules, especially in Peyer's patches, in $\alpha4^{\Delta/\Delta}$ mice. (**b**) Splenocytes were stained for expression of CD4/CD25/Foxp3, using a mouse regulatory T-cell staining kit. 10^6 cells/sample were surface-stained with a cocktail of CD4-fluorescein isothiocyanate and CD25-allophycocyanin. Subsequently, cells were fixed, permeabilized, and then stained intracellularly with Foxp3-phycoerythrin (PE) or with rat anti-mouse isotype control PE antibodies. Stained samples were analyzed by flow cytometry. Fluorescein-activated cell-sorting plots show the percentage of $CD25^+FoxP3^+$ cells gated on $CD4^+$ T cells

However, only partial hematopoietic reconstitution was present in these mixed chimeras and absolute thymic cellularity was not presented.

Our data showed decreased cellularity of thymus up to 8 months posttransplantation in the Rag 2−/− recipients of α4-deficient cells. Subset analysis suggested that the thymic hypocellularity was likely due to impaired homing of thymic progenitors and less so to their subsequent intrathymic development, as the profiles of double negative, double positive, and, single positive were not significantly different between α4f/f and α4Δ/Δ recipients (Fig. 5.4). Recent studies indicate that homing to thymus is accomplished through preferential migration of the CLP-2 population, which coexpresses P-selectin glycoprotein ligand-1, α4b1, and aLβ2 integrins and interacts with their respective ligands on thymic endothelial cells [7, 24]. This interaction, as well as the correct localization of homed cells in thymic cortex, is enforced by chemoattractant CCL25 expressed mainly in cortex and its receptor (CCR9) on CLP-2 cells. CCR9 is coexpressed with α4b1 in the latter cells and is required for homing, but had no effect on their subsequent T-cell development [25]. The latter appeared to be dependent on another chemokine, CXCL12, which did not affect thymic homing, but had a critical role in expansion and differentiation of thymocytes after their homing to thymus [26, 27]. However, in other studies, pertussis toxin treatment of cells reduced the homing of CLP-2 to thymus partially, but had no effect on CLP-2 homing to BM [7]. Because thymic homing can be subserved by P-selectin glycoprotein ligand-1 or aLβ2 as indicated by these previous studies, it may not be surprising that the homing defects of thymic progenitors were partial in the absence of α4b1 alone, or after ablation of b1 postengraftment [2], although in the latter case only transient impairment in thymic repopulation was seen and some homing to thymus might have preceded their b1 ablation, which was initiated post-full engraftment. As the total numbers of DP and single positive cells were decreased in thymus of α4Δ/Δ recipients, additional effects on their development cannot be excluded. In line with this view are antibody data implicating VLA4 in trafficking between cortex and medulla [28], thereby suggesting a nonredundant role of α4b1 integrin for intrathymic expansion/maturation of thymic progenitors [8]. Such an outcome also advocated involvement of the CXCL12/CXCR4 pathway for the intrathymic traffic and correct localization of progenitor cells for their further maturation [27]. It is of note that expansion of T cells in both fetal and adult thyme was reduced in CXCL12/CXCR4KOS [26], like in our studies. Thus, on the strength of these data, one could conclude that cooperation between the two pathways likely through inside-out signaling affecting α4b1 may be operative in this process. It is important to emphasize that the effect of α4b1, CXCL12, or aLβ2 by antibody blockade [7], like the present data, was partial, suggesting additional alternative pathways. The presence of only a very small population of CD8+ cells in the reconstituted thymus, in contrast to the normal CD4:CD8 ratio in blood is of interest. It is unlikely that it is due to the inability of CD8+ α4-deficient cells to reenter thymus [29], as this was also observed in thymi reconstituted with control cells. It was also seen previously in ES α4 null chimeras with Rag2−/− ES cells [22]. Whether some alterations in Rag2−/− microenvironments (stroma and endothelial elements [19, 30, 31]) play a role is only speculative at present. Furthermore, the detailed molecular signaling of chemokine/integrin interactions in the thymus has not been fully explored, but efforts to elucidate such molecular pathways are underway [7].

Like the reduced repopulation of thymus, severely reduced cellularity was also seen in gut-associated lymphoid cells, i.e., PPs and MLNs. As all lymphocyte subsets were similarly reduced in these tissues, the data implied a severe homing problem. A similar outcome was seen in the α4b1 antibody blockade [32, 33]. Of interest, anti-VCAM-1 [34] did not block homing to PP, unless aLβ2 was also blocked [34]. Because homing to PPs is mainly dependent on α4b7/MadCAM-1 pathway (and most of the adult CD3+ cells in these tissues are negative for b1 integrin), this was not an unexpected finding. However, PPs were reported to be normal in b7 KOS and were normal in number (but of smaller size) in b1/b7 doubly deficient mice [2]. Because the total cellularity was not assessed in the previous study with FL α4KO cells [12] or in b1/b7 double KOS, differences may be explained on methodological or technical grounds. The presence of VCAM-1+ and ICAM-1+ stroma was deemed important for PP organogenesis [35] and for colonization by CD4+/CD3− cells, a unique subset of cells expressing both α4b1 and α4b7 integrins [36]. Also of interest, MLNs provide a supportive function for DP cells, as the only such site outside thymus [37]. The presence of CCL25 in the stromal cells of both tissues may be mediating this effect. Thus, our data suggest that similar α4 integrin-dependent interactions with stromal tissue cells seem to play major roles both for homing and intrathymic development of T lymphocytes and for colonization of mucosal tissues by T-lymphoid cells.

Cellularity of all tested peripheral lymph nodes, as well as that of spleen, was similar to one found in animals repopulated with control cells (Fig. 5.2). Likewise, no abnormalities have been reported in these tissues in any of the other models [2, 12]. Nevertheless, certain abnormalities in the functional properties of mature lymphoid cells were found here and are of interest. There was a decrease in initial IgM response after immunization and antibody-specific IgE responses in mice repopulated with α4Δ/Δ cells compared to ones with α4f/f (Fig. 5.5a). A similar impairment in initial B-cell responses was noted in VCAM-1-deficient mice and in mice with ablated b1-integrin postengraftment [2, 38, 39]. Furthermore, OVA-specific IgE response after induction of allergen (OVA)-dependent acute asthma was significantly impaired when α4-deficient mice were used [20]. All these concordant data suggest weakened interactions between T helper and B cells. Whether the reduced response reflects inadequate cell contact of B cells or their subsequent expansion and stimulation is unclear, but is consistent with recent observations suggesting involvement and incorporation of α4b1 in the activation complex (pSMAC) during immune synapse formation between APCs and T cells [40].

An important mechanistic insight into the requirement of VLA4/VCAM-1 pathway for B-cell activation was recently presented. The interaction is mediated either by B-cell tethering to the target membrane and thereby facilitating BCR/antigen engagement or by enhancing the level of B-cell signaling [41]. Thus, VCAM-1, expressed on the surface of target cells (follicular dendritic cells, vascular endothelium) may capture B cells through interaction with VLA4 and enhance a mature immune synapse formation and B-cell activation [41]. Nevertheless, we found no generalized attenuation of immune responses by α4-deficient cells. B-cell proliferation to lipopolysaccharide or anti-IgM in vitro was similar to controls, but different microenvironment-dependent responses in vivo cannot be excluded. Interferon responses of stimulated T cells were also of interest. Purified naïve CD4 + CD62L + α4-deficient lymphocytes had similar IFN-γ secretion, although freshly isolated splenic CD4+ α4-deficient T cells produced less IFN-γ (Fig. 5.5b). This conflicting result may be due to contamination of CD4+ expressing cells by other cells, such as NK T cells. If NK T cells are reduced in α4-deficient mice, then we might expect to observe a reduction in IFN-γ production from freshly isolated splenocytes. This reduction would not be seen when CD4 + CD62L + cells were isolated, as NK T cells do not express CD62L. However, for definitive conclusions, further investigation is needed.

Once stimulated, α4-deficient T cells turned into IFN-γ hyperproducing cells (Fig. 5.5b), due to an increase in the number and proliferation of antigen-specific cells retained in the spleen. In addition to lack of migration out of the spleen, this result could also reflect a bias of α4-deficient lymphocytes toward Th1 responses with secondary suppression of Th2 responses. Failure to induce allergen-dependent acute asthma in our α4-deficient mice was primarily due to the inability of α4Δ/Δ lymphocyte migration to lung as well as to airways, but Th2-dependent responses in these mice were also reduced [20]. In this context, it is of interest that miR-155 was found to be critically involved in the in vivo immune response by exerting its function at the level of cytokine production, i.e., less IFN-γ but more IL-4, favoring Th2 differentiation [42]. Future studies may shed more light on this issue. In summary, the data presented herein amplify and fine-tune previous observations on the role of α4 integrins in the repopulation of lymphoid organs, by securing its role in homing to thymus and gut lymphoid tissue and by strengthening previous evidence of attenuated B-cell responses by α4-deficient cells.

Conclusion

These data complement previous observations by defending the role of α4 integrin in thymic and gut lymphoid tissue homing and by strengthening evidence of attenuated B-cell responses in α4-deficient mice.

References

1. Luster AD, Alon R, von Andrian UH. Immune cell migration in inflammation: present and future therapeutic targets. Nat Immunol. 2005;6:1182–90.
2. Brakebusch C, Bouvard D, Stanchi F, Sakai T, Fassler R. Integrins in invasive growth. J Clin Invest. 2002;109:999–1006.
3. Hynes RO. Integrins: bidirectional, allosteric signaling machines. Cell. 2002;110:673–87.
4. Berlin C, Berg EL, Briskin MJ, et al. Alphα4beta7 integrin mediates lymphocyte binding to the mucosal vascular address in MAdCAM-1. Cell. 1993;74:185–95.
5. Rose DM. The role of the alphα4 integrin-paxillin interaction in regulating leukocyte trafficking. Exp Mol Med. 2006;38:191–5.
6. Mora JR, von Andrian UH. T-cell homing specificity and plasticity: new concepts and future challenges. Trends Immunol. 2006;27:235–43.
7. Scimone ML, Aifantis I, Apostolou I, von Boehmer H, von Andrian UH. A multistep adhesion cascade for lymphoid progenitor cell homing to the thymus. PNAS. 2006;103:7006–11.
8. Schmeissner PJ, Xie H, Smilenov LB, Shu F, Marcantonio EE. Integrin functions play a key role in the differentiation of thymocytes in vivo. J Immunol. 2001;67:3715–24.
9. Halin C, Scimone ML, Bonasio R, et al. The S1P-analog FTY720 differentially modulates T-cell homing via HEV: T-cell-expressed S1P1 amplifies integrin activation in peripheral lymph nodes but not in Peyer patches. Blood. 2005;106:1314–22.
10. Arroyo AG, Yang JT, Rayburn H, Hynes RO. Differential requirements for alphα4 integrins during fetal and adult hematopoiesis. Cell. 1996;85:997–1108.
11. Bell E, Sparshott S, Ager A. Migration pathways of CD4 T cell subsets in vivo: the CD45RC- subset enters the thymus via alpha 4 integrin- VCAM-1 interaction. Int Immunol. 1995;7:1861–71.
12. Gribi R, Hook L, Ure J, Medvinsky A. The differentiation program of embryonic definitive hematopoietic stem cells is largely alphα4 integrin independent. Blood. 2006;108:501–9.
13. Bungartz G, Stiller S, Bauer M, et al. Adult murine hematopoiesis can proceed without beta1 and beta7 integrins. Blood. 2006;108:1857–64.
14. Scott L, Priestley GV, Papayannopoulou T. Deletion of alphα4 integrins from adult hematopoietic cells reveals roles in homeostasis, regeneration, and homing. Mol Cell Biol. 2003;23:9349–60.
15. Priestley GV, Scott LM, Ulyanova T, Papayannopoulou T. Lack of alphα4 integrin expression in stem cells restricts competitive function and self-renewal activity. Blood. 2006;107:2959–67.
16. Priestley GV, Ulyanova T, Papayannopoulou T. Sustained alterations in biodistribution of stem/ progenitor cells in Tie2cre + alphα4f/f mice are hematopoietic cell autonomous. Blood. 2007;109:109–11.
17. Ulyanova T, Scott L, Priestley G, et al. VCAM-1 expression in adult hematopoietic and nonhematopoietic cells is controlled by tissue inductive signals and reflects their developmental origin. Blood. 2005;106:86–94.
18. McAdam AJ, Greenwald RJ, Levin MA, et al. ICOS is critical for CD40-mediated antibody class switching. Nature. 2001;409:102–5.
19. Prockop SE, Petrie HT. Regulation of thymus size by competition for stromal niches among early t cell progenitors. J Immunol. 2004;173:1604–11.
20. Banerjee ER, Jiang Y, Henderson JWR, Scott LM, Papayannopoulou T. alphα4 and beta2 integrins have nonredundant roles for asthma development, but for optimal allergen sensitization only alphα4 is critical. Exp Hematol. 2007;35:605–17.
21. Issekutz T. Inhibition of in vivo lymphocyte migration to inflammation and homing to lymphoid tissues by the TA-2 monoclonal antibody. A likely role for VLA-4 in vivo. J Immunol. 1991;147:4178–84.
22. Arroyo AG, Taverna D, Whittaker CA, et al. In vivo roles of integrins during leukocyte development and traffic: insights from the analysis of mice chimeric for alpha5, alphav, and alphα4 integrins. J Immunol. 2000;165:4667–75.
23. Arroyo AG, Yang JT, Rayburn H, Hynes RO. Alphα4 integrins regulate the proliferation/differentiation balance of multilineage hematopoietic progenitors in vivo. Immunity. 1999;11:555.
24. Rossi FMV, Corbel SY, Merzaban JS, et al. Recruitment of adult thymic progenitors is regulated by P-selectin and its ligand PSGL-1. Nat Immunol. 2005;6:626–34.
25. Parmo-Cabanas M, Garcia-Bernal D, Garcia-Verdugo R, Kremer L, Marquez G, Teixido J. Intracellular signaling required for CCL25- stimulated T cell adhesion mediated by the integrin alphα4 beta1. J Leukoc Biol. 2007;82:380–91.
26. Ara T, Itoi M, Kawabata K, et al. A role of CXC chemokine ligand 12/stromal cell-derived factor-1/pre-B cell growth stimulating factor and its receptor CXCR4 in fetal and adult T cell development in vivo. J Immunol. 2003;170:4649–55.
27. Plotkin J, Prockop SE, Lepique A, Petrie HT. Critical role for CXCR4 signaling in progenitor localization and T cell differentiation in the postnatal thymus. J Immunol. 2003;171:4521–7.
28. Crisa L, Cirulli V, Ellisman MH, et al. Cell adhesion and migration are regulated at distinct stages of thymic T cell development: the roles of fibronectin, VLA4, and VLA5. J Exp Med. 1996;184:215–28.
29. Sitnicka E, Buza-Vidas N, Ahlenius H, et al. Critical role of FLT3 = ligand in IL-7 receptor-independent T lymphopoiesis and regulation of lymphoid-primed multipotent progenitors. Blood. 2007;110: 2955–64.
30. Boyd RL, Tucek CL, Godfrey DI, et al. The thymic microenvironment. Immunol Today. 1993;14:445–59.

31. van Ewijk W, Shores EW, Singer A. Crosstalk in the mouse thymus. Immunol Today. 1994;15: 214–17.
32. Hamann A, Andrew DP, Jablonski-Westrich D, Holzmann B, Butcher EC. Role of alpha 4-integrins in lymphocyte homing to mucosal tissues in vivo. J Immunol. 1994;152:3282–93.
33. Xu B, Wagner N, Pham LN, et al. Lymphocyte homing to bronchus associated lymphoid tissue (BALT) is mediated by L-selectin/PNAd, alphα4beta1 integrin/VCAM-1, and LFA-1 adhesion pathways. J Exp Med. 2003;197:1255–67.
34. Berlin-Rufenach C, Otto F, Mathies M, et al. Lymphocyte migration in lymphocyte function-associated antigen (LFA)-1-deficient mice. J Exp Med. 1999;189:1467–78.
35. Finke D, Kraehenbuhl J-P. Formation of Peyer's patches. Curr Opin Genet Dev. 2001;11:561–7.
36. Finke D, Acha-Orbea H, Mattis A, Lipp M, Kraehenbuhl JP. CD4 + CD3- cells induce Peyer's patch development: role of alphα4- beta1 integrin activation by CXCR5. Immunity. 2002;17:363–73.
37. Maillard I, Schwarz BA, Sambandam A, et al. Notch-dependent T-lineage commitment occurs at extrathymic sites following bone marrow transplantation. Blood. 2006;107:3511–19.
38. Koni PA, Joshi SK, Temann U-A, Olson D, Burkly L, Flavell RA. Conditional vascular cell adhesion molecule 1 deletion in mice: impaired lymphocyte migration to bone marrow. J Exp Med. 2001;193:741–54.
39. Leuker CE, Labow M, Muller W, Wagner N. Neonatally induced inactivation of the vascular cell adhesion molecule 1 gene impairs B cell localization and T cell-dependent humoral immune response. J Exp Med. 2001;193:755–68.
40. Barreiro O, de la Fuente H, Mittelbrunn M, Sanchez-Madrid F. Functional insights on the polarized redistribution of leukocyte integrins and their ligands during leukocyte migration and immune interactions. Immunol Rev. 2007;218:147–64.
41. Carrasco YR, Batista FD. B-cell activation by membrane-bound antigens is facilitated by the interaction of VLA-4 with VCAM-1. EMBO J. 2006;25:889–99.
42. Thai T-H, Calado DP, Casola S, et al. Regulation of the germinal center response by microRNA-155. Science. 2007;316:604–8.

Published in

Banerjee ER, Latchman YL, Jiang Y, Priestley GV, Papayannopoulou T. Distinct changes in adult lymphopoiesis in Rag2–/– mice fully reconstituted by α4-deficinet adult bone marrow cells. Exp Hematol. 2008;36(8):1004–13.

Studying the Roles of Some Critical Molecules in Systemic Inflammation

6

Abstract

Objective. Leukocyte recruitment to inflammatory sites is a prominent feature of acute and chronic inflammation. Instrumental in this process is the coordinated upregulation of leukocyte integrins (among which α4b1 and β2 integrins are major players) and their cognate receptors in inflamed tissues. To avoid the ambiguity of previous short-term antibody-based studies and to allow for long-term observation, we used genetically deficient mice to compare roles of α4 and β2 integrins in leukocyte trafficking.

Materials and Methods . Aseptic peritonitis was induced in α4 or β2 integrin-deficient (conditional and conventional knockouts, respectively) and control mice, and recruitment of major leukocyte subsets to the inflamed peritoneum was followed for up to 4 days.

Results. Despite normal chemokine levels in the peritoneum and adequate numbers, optimal recruitment of myeloid cells was impaired in both α4- and β2-deficient mice. Furthermore, clearance of recruited neutrophils and macrophages was delayed in these mice. Lymphocyte migration to the peritoneum in the absence of α4 integrins was drastically decreased, both at steady state and during inflammation, a finding consistent with impaired lymphocyte in vitro adhesion and signaling. By contrast, in the absence of β2 integrins, defects in lymphocyte recruitment were only evident when peritonitis was established.

Conclusions. Our data with concurrent use of genetic models of integrin deficiency reveal nonredundant functions of α4 integrins in lymphocyte migration to the peritoneum and further refine specific roles of α4 and β2 integrins concerning trafficking and clearance of other leukocyte subsets at homeostasis and during inflammation.

E.R. Banerjee, *Perspectives in Inflammation Biology*,
DOI 10.1007/978-81-322-1578-3_6, © Springer India 2014

Unique and Redundant Roles of α4 and β2 Integrins in Kinetics of Recruitment of Lymphoid Versus Myeloid Cell Subsets to the Inflamed Peritoneum Revealed by Studies of Genetically Deficient Mice

Introduction

A prominent feature of acute or chronic inflammation is the recruitment of mature leukocytes to inflammatory sites. For the successful implementation of this process, several highly coordinated adhesion and activation steps need to be accomplished by leukocytes in inflamed tissues [1, 2]. Essential molecular players in this multistep adhesion/migration cascade are α4 and β2 integrins. In particular, α4b1 (VLA4) integrin is unique among integrins as it can function in all three steps of the trafficking cascade: rolling/tethering initiated by selectins, firm adhesion, and transmigration controlled by activated integrins [3–6]. Expression of α4 integrins is constitutive in all leukocytes except human neutrophils, where it is inducible [7], whereas murine neutrophils constitutively express α4b1 [8]. The β2 integrins are expressed exclusively in hematopoietic cells [9]. Both α4 and β2 integrins, as well as their cognate receptors, are upregulated by various inflammatory stimuli [1, 6]. Function-blocking antibodies and peptides have been extensively used to study the role of α4 and β2 integrins in leukocyte trafficking. However, results of antibody studies vary with the animal model used or the route of antibody administration, and off-target effects cannot be excluded [10–14]. To avoid the ambiguity of antibody-based studies and to carry out long-term observations, mouse models with genetically modified integrin genes have been generated [15–17]. To circumvent embryonic lethality of α4 knockout mice [18] in studying the role of α4 integrins in vivo, reconstitution of RAG_/_ mice with α4-deficient embryonic stem cells was undertaken [19]. In this model though, a profound defect in development of α4_/_ lymphocytes and lymphoid organs was observed in postnatal life, thus precluding the use of this model to study migratory behavior of mature leukocyte populations. A new model of postnatal conditional α4 deficiency with normal development of the immune system was recently established in our laboratory [20]. Using this model, new aspects of the role of α4 integrins in homing and retention of hematopoietic progenitors in the bone marrow at steady state and recovery after hematopoietic stress were revealed, but trafficking patterns of mature leukocytes to the inflammatory sites have not previously been addressed or compared to other integrin-deficient mice. To uncover unique and overlapping roles of α4 and β2 integrins in mature hematopoietic cell trafficking, we analyzed patterns of recruitment of various leukocyte subsets to the peritoneum before and after inflammation in α4 or β2 integrin-deficient mice, using a well-studied model of aseptic thioglycolate-induced peritonitis. Our results revealed intrinsic differences of migratory responses in the absence of α4 integrins in lymphoid versus myeloid subsets. Parallel studies using mice with single α4 or β2, as well as mice with double (α4 and β2) integrin deficiency allowed fine-tuning of the roles of α4 and β2 integrins in leukocyte trafficking.

Materials and Methods

Mice

Mice used in this study were of C57/Bl6x129 (WT, α4Δ/Δ) or C57/Bl6 (CD18–/–) background, between 8 and 12 weeks of age. Wild-type (WT) animals were purchased from Taconic (Germantown, NY, USA). β2 integrin-deficient mice were obtained from Dr. A. Beaudet (Baylor College, Houston, TX, USA) [16]. M_Cre + α4f/f mice were generated in our laboratory [20]. To induce α4 integrin ablation, these mice were treated neonatally with interferon inducer, poly(I:C) (three injections of 50 mL of 1 mg/mL solution in phosphate-buffered saline (PBS), intraperitoneally (48 h apart)). Metalloproteinases (MMP)-9_/_ mice were kindly provided by Dr. R. Senior (Washington University, St. Louis, MO, USA) [21]. All animals were bred and maintained under specific pathogen-free conditions at the University of Washington.

All experimental procedures were done in accordance with Institutional Animal Care and Use Committee guidelines on approved protocols.

Antibodies

Anti-α4 integrin antibody, PS/2, was purchased from Southern Biotechnology (Birmingham, AL, USA). Phycoerythrin (PE)-Cy5-conjugated F4/80 (Cl:A3-1) was from AbD Serotec Ltd (Raleigh, NC, USA). Gr 1-PE, Gr1-allophycocyanin (APC), CD45-APC, CD45-fluorescein isothiocyanate (FITC), CD3-Cy-chrome, CD4-Cy-Chrome, CD8-FITC, B220-FITC, B220-PE, β220-Cy-Chrome, CD19-biotin, CD18-PE, streptavidin-APC, and corresponding fluorochrome-conjugated isotype-matched immunoglobulins that served as controls were purchased from BD Biosciences (San Diego, CA, USA).

Peritoneal Inflammation

Mice were injected intraperitoneally with 1 mL 3 % thioglycolate (TG) and were sacrificed by cervical dislocation 4, 16, 48, and 96 h postinjection. The peritoneal cavity was lavaged with 5 mL ice-cold PBS containing 5 mM ethylenediamine tetraacetic acid; cells were enumerated and used for further analysis. Chemokine measurement aliquots of frozen peritoneal lavage fluid were sent to Pierce Biotechnology Inc. (Woburn, MA, USA) to measure concentration of an array of chemokines and cytokines by Searchlight Multiplex technology.

Fluorescein-Activated Cell-Sorting Analysis

Antibody-labeled cells were analyzed on a FACSCalibur (BD Biosciences, San Jose, CA, USA) using CellQuest software. In vitro migration trans-well migration was performed as described elsewhere [22]. In brief, splenocytes were stained with B220-FITC and CD3-PE, and 0.5×10^6 cells were transferred into the upper chambers of trans-well inserts (pore size 5 mm, Corning Costar, Cambridge, MA, USA). Splenocytes were allowed to migrate through the uncoated inserts (trans-well migration) or trans-wells coated with bEND3 mouse endothelial cells (transendothelial migration) for 4 h at 37_C toward stromal-derived factor-1a (SDF-1a; 100 ng/mL, Peprotech, Rocky Hill, NJ, USA). Cells collected from the lower chamber were enumerated by fluorescein-activated cell sorting (FACS). The number of events acquired from this sample was compared to that of input cells (before migration), and the percentage of migration was calculated. Splenocytes were isolated and labeled with carboxyfluorescein succinimidyl ester (Invitrogen, Carlsbad, CA, USA) according to the manufacturer's instructions, washed and brought to 2×10^6 cells/mL in adhesion buffer (PBS containing 1 % bovine serum albumin, 2 mM MgCl2, 2 mM CaCl2). Labeled cells were transferred into 24-well plates (1×10^6 cells/well) with a confluent layer of bEND3 cells either untreated or treated overnight with tumor necrosis factor-a (TNF-α; 100 ng/mL, Peprotech). After a short spin (2 min, 100 g), adhesion was allowed to occur at 37 °C for 40 min. Nonadherent cells were washed out with PBS. bEND3 cells and adherent splenocytes were trypsinized with TrypLE Select (Invitrogen) and washed, and the number of carboxyfluorescein succinimidyl ester-positive cells was determined by FACS after 1-min acquisition. Cell adhesion was expressed as a percent of input cells.

Actin Polymerization

Actin polymerization was performed as described previously [23]. In brief, splenocytes (1×10^6 cells) were stained with CD3-FITC and B220-FITC, stimulated with 100 ng/mL SDF-1α (Peprotech) for indicated amount of time, fixed, permeabilized, and stained with phalloidin (Invitrogen). After washing with PBS/bovine serum albumin, cells, gated on lymphocytes (CD3+ B220+), were analyzed by FACS.

Ca^2+ Mobilization

Ca^2+ mobilization experiments were performed as described elsewhere [24]. Splenocytes were isolated and loaded with Indo-1 (Invitrogen), washed, and labeled with β220-FITC and CD3-PE antibodies. Cells were resuspended in Ca^2+ buffer (PBS, 1 % bovine serum albumin, 1 mM CaCl2, 1 mM MgCl2) at 4×10^6 cells/mL. Change of the 395/530 fluorescence quotient was monitored over 3 min by FACS following stimulation with SDF-1α (200 ng/mL, Peprotech) or Ionomycin (1 mM, Sigma). After analysis using FloJo software, Ca^2+ mobilization was expressed as a percent of the peak levels after stimulation over baseline.

Statistical Analysis

Data shown are mean six standard error of mean. Statistical analyses were performed using a Student's *t*-test and $p < 0.05$ was considered significant.

Results in a Nutshell. Despite normal chemokine levels in the peritoneum and adequate numbers, optimal recruitment of myeloid cells was impaired in both α4- and β2-deficient mice. Furthermore, clearance of recruited neutrophils and macrophages was delayed in these mice. Lymphocyte migration to the peritoneum in the absence of α4 integrins was drastically decreased, both at steady state and during inflammation, a finding consistent with impaired lymphocyte in vitro adhesion and signaling. By contrast, in the absence of β2 integrins, defects in lymphocyte recruitment were only evident when peritonitis was established.

Results

Animal Models

In the present study, we used genetic models of α4 (α4b1 and α4b7) or β2 (CD18) integrin deficiency. β2 integrin-deficient (CD18_/_) mice were generated by germline deletion of β2 integrin gene [16]. α4 Integrin-deficient mice (α4Δ/Δ) were previously generated in our laboratory as conditional knockouts. These were MxCre + α4f/f mice, in which ablation of α4 integrins occurs after treatment with interferon inducer, poly(I:C) [20]. After ablation of α4 integrin in these mice, percent of α4+ cells was reduced to 3.7 % 6 0.2 % (n 5 11) in the bone marrow and 5.2 % 6 0.8 % (*n* 5 11) in the peripheral blood. White blood cell (WBC) counts in both α4 integrin-deficient ($19 \pm 2 \times 10^3$ cells/mL) and β2 integrin-deficient ($49 \pm 5 \times 10^3$ cells/mL) mice were significantly higher than in control mice ($8 \pm 1 \times 10^3$ cells/mL), as described previously [16, 20]. In addition, we recently generated mice with both α4 and β2 integrin deficiency by interbreeding single integrin knockouts. Here we report only preliminary results with α4Δ/Δ CD18–/– doubly deficient mice, since only a handful of animals were available for experiments due to difficulties in breeding.

To Study Unique and Redundant Roles of α4 and β2 Integrins

In leukocyte migration in vivo, we employed the aseptic peritonitis model in all mice with integrin deficiencies and monitored various leukocyte subsets by comparing their levels in circulation and in the peritoneal cavity for intervals up to 4 days.

Kinetics of Migration of Various Leukocyte Subsets In Vivo

Neutrophils are the first leukocytes that migrate and accumulate at a site of inflammation. By the 4th hour of peritonitis, a substantial number of neutrophils was detected in the peritoneum of WT and α4Δ/Δ mice and increased further by the 16th hour (Fig. 6.1a). In CD18–/– mice, neutrophils were slower to migrate, but by the 16th hour of peritonitis their numbers were comparable to those in WT animals (Fig. 6.1a). Defects in optimal neutrophil recruitment became

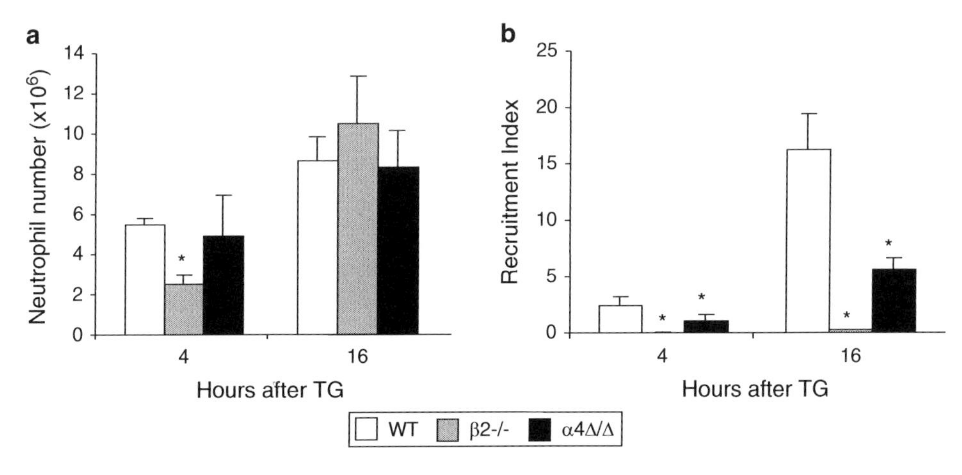

Fig. 6.1 Both α4 and β2 integrins are required for efficient neutrophil migration to the peritoneum. Mice were injected with thioglycolate (TG), peritoneal leukocytes were harvested 4 and 16 h postinjection and stained with various leukocyte markers (see Materials and Methods). Neutrophil (Gr-1^+, F4/80^-) numbers (**a**) and recruitment indices (**b**) were calculated. Note that although sufficient numbers of neutrophils are present in the peritoneal cavity of α4Δ/Δ and $\beta2^{-/-}$ mice, there is a significant decrease in recruitment index at each time point. *Significant difference compared to controls, $p < 0.05$. Mice used per group: wild type (WT), $n = 5$; $\beta2^{-/-}$, $n = 4$; α4Δ/Δ, $n = 4$

evident when we compared recruitment indices in α4 and β2 integrin-deficient versus control mice (Fig. 6.1b). Recruitment index represents the proportion of cells harvested from the peritoneal cavity measured against the total number of cells available for migration in the circulation, assuming a blood volume of 2 mL in the mouse. The defect was more pronounced in CD18–/– mice: recruitment indices at each time point were < 2 % of corresponding control value, whereas in α4Δ/Δ mice at 4 and 16 h, they comprised 46 % and 34 %, respectively, of the corresponding control value. This observed defect in CD18–/– neutrophil recruitment was even more striking, in view of elevated levels of neutrophil chemoattractant MIP2 in the peritoneal cavity of CD18–/– mice (267 6 63 pg/mL) as compared to controls (74 6 5 pg/mL, $p < 0.05$). In the α4-deficient mice, MIP2 levels were similar to controls, as were KC levels in both α4_/_ and CD18–/– mice (data not shown). At later times of peritonitis, (96 h), more neutrophils were harvested from the peritoneum of both α4 ($0.4 \pm 0.03 \times 10^6$ cells, $p < 0.01$) and β2 ($3.7 \pm 0.9 \times 10^6$ cells, $p < 0.01$) integrin-deficient mice as compared to controls ($0.1 \pm 0.02 \times 10^6$ cells). It is worth noting that, in WT mice, all peritoneal neutrophils were α4 positive, whereas in α4-deficient mice, all migrated neutrophils were α4 negative (data not shown), suggesting their α4-independent migration. As a general rule, macrophage recruitment follows that of neutrophils during inflammation. We first analyzed inflammatory monocytes, a subset that is recognized by the presence of Gr-1 marker in addition to the monocytic marker, F4/80 [25] (Fig. 6.2a). Although there was no difference in the numbers of inflammatory monocytes at 4 h, at the peak of their accumulation (16 h), significantly fewer inflammatory monocytes were recovered from peritoneum of either α4 or β2 integrin-deficient mice than in controls (Fig. 6.2b), despite their increased numbers in circulation (Fig. 6.2c). Calculated recruitment indices further emphasized impairment in recruitment of inflammatory monocytes in integrin-deficient mice (Fig. 6.2d). The observed defect occurred despite normal levels of MCP1 and MIP1a, mononuclear cell chemokines, in the peritoneal cavity of these mice (data not shown). We next studied accumulation of total peritoneal macrophages (F4/80-positive cells). The numbers of F4/80+ cells recovered from the peritoneum were similar in integrin- deficient and control mice, except for a transient but significant

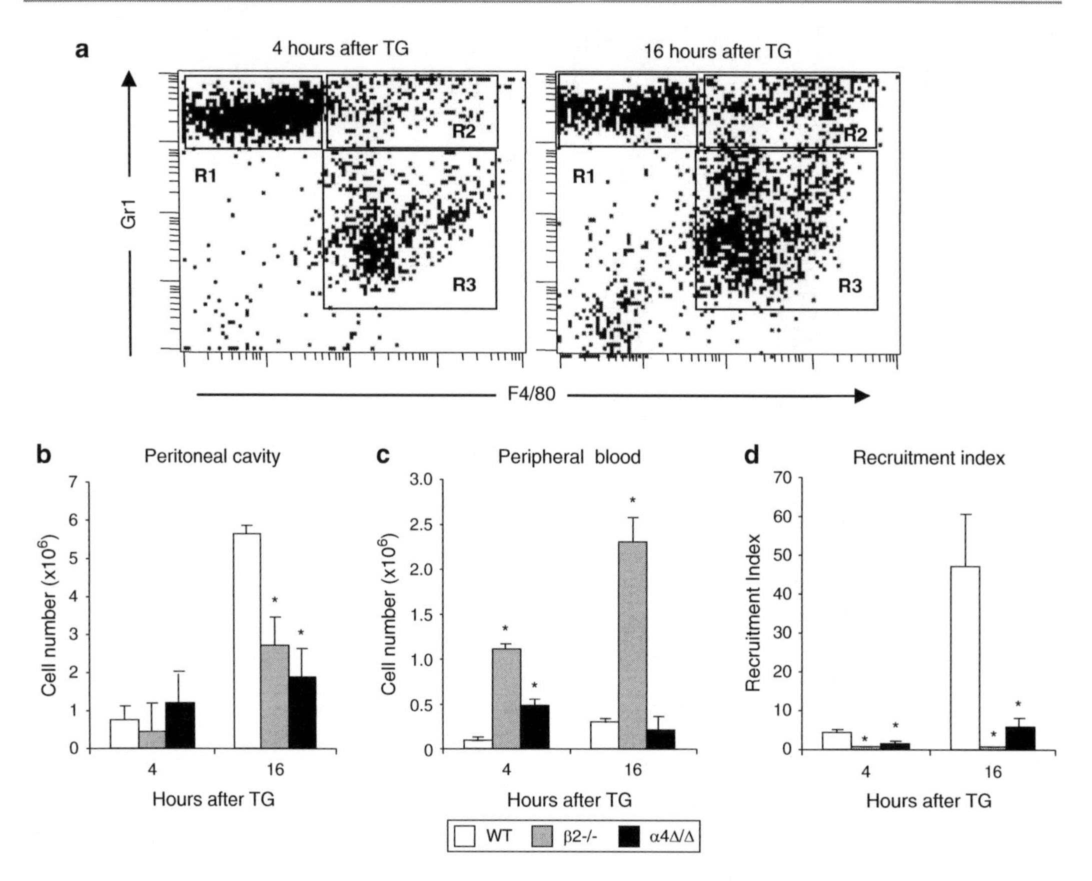

Fig. 6.2 Recruitment of inflammatory monocytes to the peritoneum is impaired in α4Δ/Δ and $\beta2^{-/-}$ mice. Aseptic peritonitis was induced in mice by thioglycolate (TG) injection and 4 and 16 h later, numbers of inflammatory monocytes were counted. (**a**) Typical fluorescein-activated cells sorting profile of peritoneal exudate of a wild-type (WT) mouse 4 h (*left panel*) and 16 h (*right panel*) after TG injection. Neutrophils are Gr-1^{+}F4/80^{-} (R1), inflammatory monocytes are Gr-1^{+}F4/80^{+} (R2), and non-inflammatory monocytes/macrophages are Gr-1^{-}F4/80^{+} (R3); numbers of inflammatory monocytes in WT ($n=5$), $\beta2^{-/-}$ ($n=4$), and α4Δ/Δ ($n=4$) mice were determined in the peritoneal cavity (**b**) and in peripheral blood (**c**). Recruitment indices (**d**) were calculated to assess the efficiency of migration from the circulation. *Significant difference over control ($p<0.05$)

decrease in macrophage migration at 16 h in α4-deficient mice (Fig. 6.3a). Of interest, at steady state, 37.61 % ± 3.42 % of peritoneal F4/80+ macrophages were α4 positive, and this proportion gradually decreased after TG injection to 32.71 % ± 5.99 % at 4 h, 13.16 % ± 3.98 % at 16 h, and further to 2.78 % ± 1.88 % at 96 h. While the decrease in proportion of α4-positive macrophages was likely attributable in part to dilution by influxing α4-negative macrophages, their total numbers also decreased with time (Fig. 6.3b) in α4-deficient mice. To investigate recruitment of lymphocytes, we compared numbers of T and B cells in the peritoneal lavage of naïve mice and 48 h after TG injection, at which time the peak of lymphocyte recruitment is observed [26]. In CD18−/− mice, although the lymphocyte numbers in peritoneal cavity ($2.05\times10^{6}\pm0.45\times10^{6}$ cells) were similar to controls ($1.15\times10^{6}\pm0.19\times10^{6}$ cells) at steady state, a defect in lymphocyte recruitment became evident after inflammation was established: TG injection resulted in less than twofold increase over baseline in CD18−/− lymphocyte numbers that is significantly less than a sevenfold increase in controls (Fig. 6.4a), and the recruitment index was also significantly decreased (Fig. 6.4b). In nonstimulated α4-deficient animals, even though

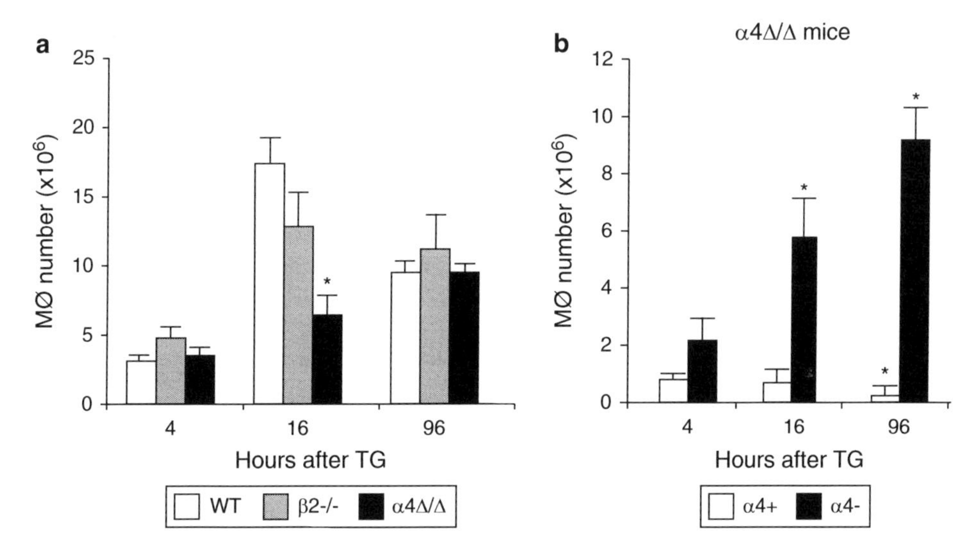

Fig. 6.3 Macrophages use α4 and β2 integrins interchangeably to maintain sufficient numbers in peritoneum during inflammation. (**a**) Peritonitis was induced in wild-type (WT) ($n=5$), $\beta2^{-/-}$ ($n=4$), and α4Δ/Δ ($n=4$) mice and 4, 16, and 96 h later numbers of macrophages ($F4/80^{+}$) and (**b**) numbers of α4-positive and -negative macrophages in α4Δ/Δ peritoneal lavage were determined by fluorescein-activated cells sorting. *Significant difference over control ($p < 0.05$); MØ 5 macrophages

only a small number was recovered from the peritoneal cavity, 47.1 % 6 7.6 % of T cells and 71.4 % 6 8.1 % of B cells were α4-positive, in contrast to 5.2 % 6 0.8 % of α4-positive cells present in the peripheral blood. These data suggest preferential migration and accumulation over time of α4-positive cells at steady state. Very few α4-negative lymphocytes were detected in peritoneal cavities of α4-deficient mice, and this number did not change after TG injection (Fig. 6.4a). The recruitment index for α4_/_ lymphocytes was 45-fold less than for control (Fig. 6.4b), indicating a profound defect in α4_/_ lymphocyte migration and recruitment. Of interest, a decrease in levels of lymphocyte chemoattractant RANTES in the peritoneal cavity was observed in α4-deficient relative to control mice (6.37 6 1.42 pg/mL and 14.63 6 1.65 pg/mL, respectively, p < 0.05) whereas measured RANTES levels in CD18−/− mice were similar to controls (data not shown).

The majority of migrated lymphocytes in all TG-induced peritonitis mouse models studied were comprised of B cells (data not shown).

In the WT mice, TG injection resulted in transient leucopenia: at 4 and 16 h of peritonitis circulating WBC numbers dropped to $3.2 \pm 0.5 \times 10^3$ and $4.0 \pm 0.5 \times 10^3$ cells/mL blood, respectively, but returned to basal levels by the 96th hour ($7.2 \pm 0.4 \times 10^3$ cells/mL blood). In contrast, in the integrin-deficient mice, no significant changes in WBC counts were detected over the course of peritonitis as compared to baseline (data not shown).

We recently developed animals with both α4 and β2 integrin deficiency by interbreeding the single integrin-deficient mice. The baseline WBC counts ($109 \pm 10 \times 10^3$ cells/mL, $n = 14$) reached unprecedented levels in these doubly deficient mice, likely reflecting not only persistent inflammatory conditions but also profound defects in leukocyte emigration to tissues. Indeed, peritoneal leukocyte numbers at 0, 4, 16, and 96 h after TG injection were $5.6 \pm 0.3 \times 10^6$ (n 5 4), $4.4 \pm 2.1 \times 10^6$ (n 5 = 4), $5.9 \pm 2.6 \times 10^6$ ($n = 3$), and $5.4 \pm 2.4 \times 10^6$ ($n = 3$) cells, respectively, indicating abrogation of leukocyte recruitment in the absence of both integrins, despite markedly elevated circulating WBCs (Fig. 6.5). Comparison of leukocyte recruitment in single integrin- and double-deficient animals indicates that both integrins

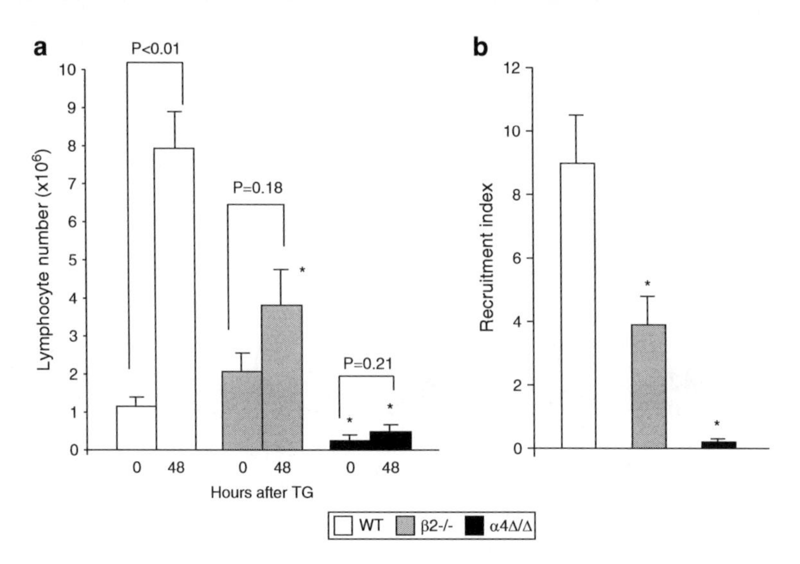

Fig. 6.4 Lymphocyte migration to the inflamed peritoneum is absent in α4Δ/Δ and decreased in $\beta 2^{-/-}$ mice. (**a**) Aseptic peritonitis was induced in wild-type (WT) (n 5 5), $\beta 2^{-/-}$ ($n = 5$), and α4Δ/Δ ($n = 5$) mice. Lymphocyte numbers were determined by fluorescein-activated cells sorting, before and 48 h after thioglycolate (TG) injection. (**b**) Efficiency of migration, as indicated by recruitment index, was calculated at 48 h after TG injection. *Significant difference from control mice at the corresponding hour of peritonitis ($p < 0.05$)

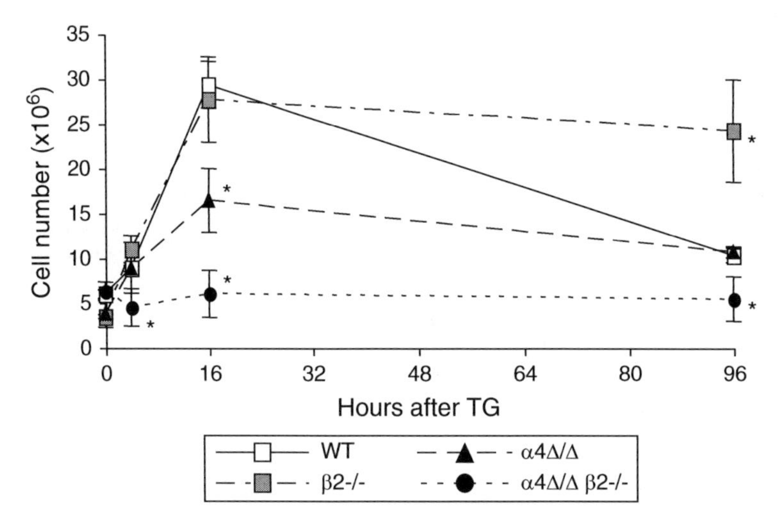

Fig. 6.5 α4 and β2 integrins are necessary and sufficient for leukocyte migration to the inflamed peritoneum. Total leukocyte numbers in peritoneum of wild type (WT) ($n = 5$), $\beta 2^{-/-}$ ($n = 5$), α4Δ/Δ ($n = 5$), and $\alpha 4\Delta/\Delta\beta 2^{-/-}$ double deficient ($n = 4$) mice at various times after induction of aseptic peritonitis. Note that, most of the time leukocyte numbers in mice deficient in a single integrin are similar to that of controls, whereas no recruitment of double-deficient leukocytes occurs. *Significant difference from control ($p < 0.05$)

are absolutely required for leukocyte migration to the peritoneal cavity.

Considering that the most prominent defect in recruitment to the inflamed peritoneum involved the α4Δ/Δ lymphocytes, we further investigated their migratory behavior in vitro. We compared adhesion of control and α4Δ/Δ lymphocytes to TNF-α-stimulated (vascular cell adhesion molecule-1 [VCAM-1] expressing) and nonstimulated bEND3 cells, a mouse endothelial cell line. Twice as many control lymphocytes adhered to VCAM-1-expressing bEND3 cells as compared to nonstimulated cells; at the same time, no increase in adhesion was observed with α4Δ/Δ lymphocytes (Fig. 6.6a). These data suggest the preferential usage of α4 integrins by lymphocytes for adhesion to bEND3 cells, since TNF-α not only upregulates VCAM-1 but also ICAM-1, a β2 integrin ligand on endothelial cells [27, 28].

To determine whether ability to migrate per se is preserved in α4Δ/Δ lymphocytes, we performed migration through an uncoated trans-well insert (trans-well migration) not requiring engagement of α4 integrins. As a stimulus for migration, we used SDF-1a, a well-characterized lymphocyte chemoattractive agent [29]. In this setting, since equal amounts of chemoattractant are present, differences in migration would reflect cell intrinsic defects. Substantial migration of B and T lymphocytes in all genotypes except for α4Δ/Δ B cells was observed (Fig. 6.6b and data not shown), suggesting impairment of signaling in these cells. We next performed lymphocyte migration through a trans-well coated with bEND3 cells (transendothelial migration). Lymphocytes were allowed to migrate through nonstimulated bEND3 cells or cells stimulated with TNF-α, upregulating VCAM-1. Control and CD18−/− B cells migrated more efficiently through TNF-α-treated endothelium; in contrast, α4Δ/Δ B cells showed no increase in migration through VCAM-1-expressing bEND3 cells (Fig. 6.6c). Similar results were seen with T cells (data not shown). Of interest, significantly fewer CD18−/− B cells migrated through the nonstimulated endothelial layer, whereas their migration through VCAM-1-expressing bEND3 cells was similar to control indicating that migration through the TNFα-stimulated endothelial cells is superseded by α4 integrins via VLA4/VCAM-1 interaction.

Since trans-well migration was impaired in α4Δ/Δ lymphocytes, we tested early signaling events characteristic of this process. Actin polymerization is a very early event in migration-associated signaling in response to SDF-1α. No significant differences in actin polymerization, measured by phalloidin staining, were detected in either integrin-deficient lymphocytes or control cells, suggesting preservation of proximal signaling steps leading to actin remodeling (Fig. 6.7a). At the same time, absence of α4 integrin in lymphocytes resulted in a modest but significant decrease in SDF-1a-induced Ca2+ mobilization (Fig. 6.7b), likely reflecting impaired signaling. These data suggest impaired response of α4-deficient lymphocytes to soluble chemoattractants. usage of α4 or β2 integrins by neutrophils may be tissue specific. While neutrophil migration to the heart and lung tissues is mediated by both α4 and β2 integrins [30, 31], migration of neutrophils to the skin in nonallergic inflammation depends primarily on β2 integrins [17, 32]. Tissue-specific variations in chemokine secretion and expression of adhesion molecules in local inflammatory milieu may explain different patterns of integrin usage by leukocytes during their migration to the various tissues [33].

Of interest are observations regarding the recruitment of monocyte/macrophages in α4-deficient mice. Macrophage numbers in these mice, although low in early development of peritonitis (Fig. 6.3a) due to delayed recruitment of inflammatory monocytes to the site of inflammation (Fig. 6.2), continually increased up to 96 h (Fig. 6.3b). As the number of α4-negative macrophages increased, the number of α4-positive macrophages progressively decreased during the course of inflammation (Fig. 6.3b). The decline in total numbers did not simply reflect “dilution” by the incoming α4-negative cells, but is best explained by their egress from the peritoneal cavity, as was previously shown for normal macrophages [34]. Macrophage egress from the peritoneum observed in control mice (compare the drop in macrophage numbers between 16 and 96 h of peritonitis, Fig. 6.3a) was impaired in both α4 and β2 integrin-deficient mice, suggesting that the presence of α4 integrins, along with the β2-integrin Mac1 [35], is important in their egress from inflammatory sites.

In contrast to macrophages which migrate from the peritoneal cavity into the draining lymphatics [34], neutrophils undergo apoptosis and are cleared from the peritoneum by macrophages. Thus, increased neutrophil numbers observed in both α4Δ/Δ and CD18−/−

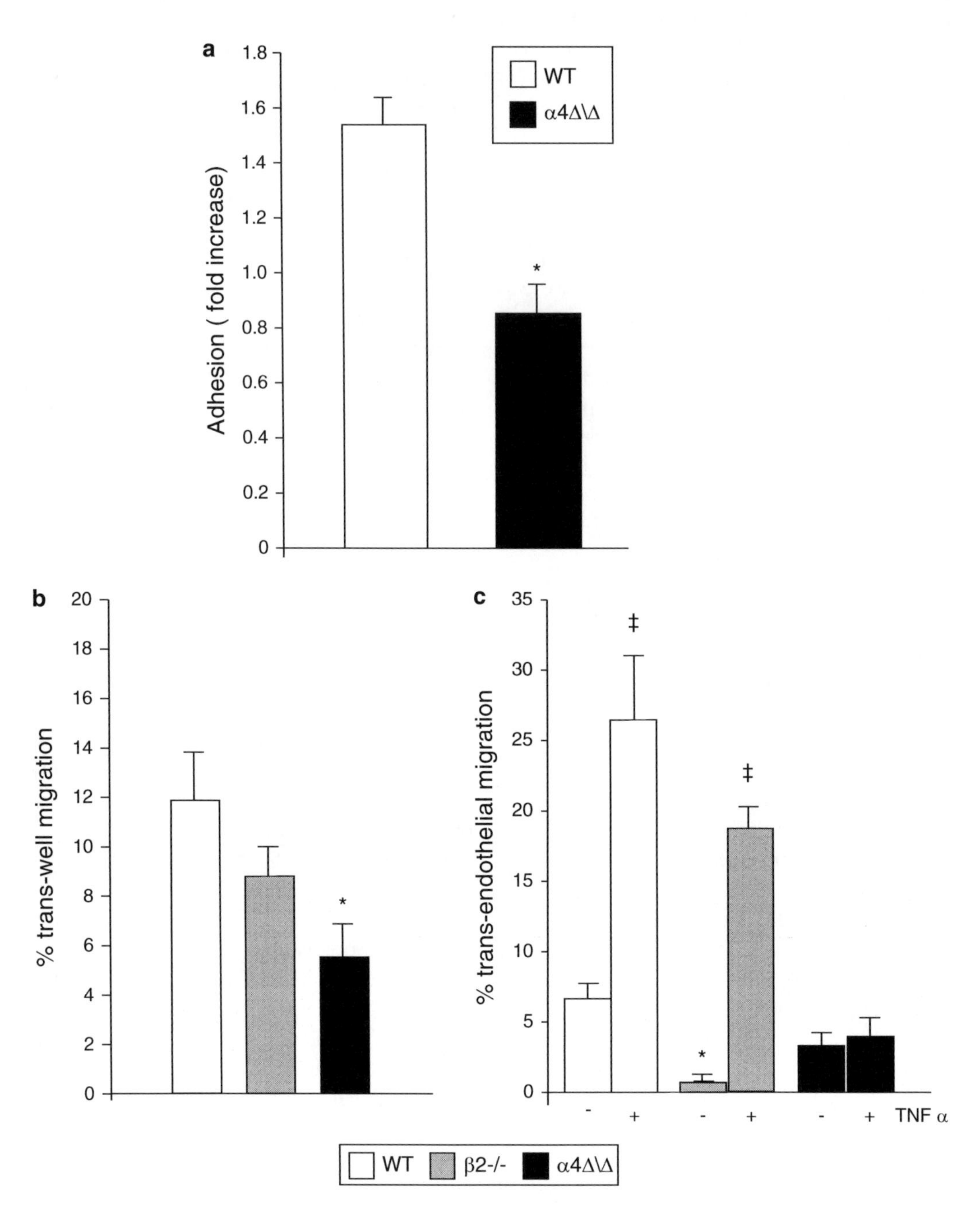

Fig. 6.6 α4 integrins are essential for lymphocyte adhesion and migration in vitro. (**a**) Adhesion of control ($n=5$) and α4Δ/Δ ($n=3$) splenocytes to endothelial cells. Splenocytes were labeled with carboxyfluorescein succinimidyl ester and allowed to adhere to either nonstimulated or tumor necrosis factor-α (TNF-α)–stimulated bEND3 monolayer. Fold increase of adhesion to stimulated over nonstimulated endothelium is displayed. (**b**) Trans-well and (**c**) transendothelial migration of B lymphocytes toward stromal-derived factor-1α (SDF-1α). Note that migration of α4Δ/Δ B cells (and T cells, not shown) through vascular cell adhesion molecule-1–expressing endothelium (TNF-α) did not differ from that of nonstimulated endothelium, in contrast to control and $\beta 2^{-/-}$ cells. *Significant difference over controls; ‡significant difference over nonstimulated endothelium ($p<0.05$). Three mice per group were used

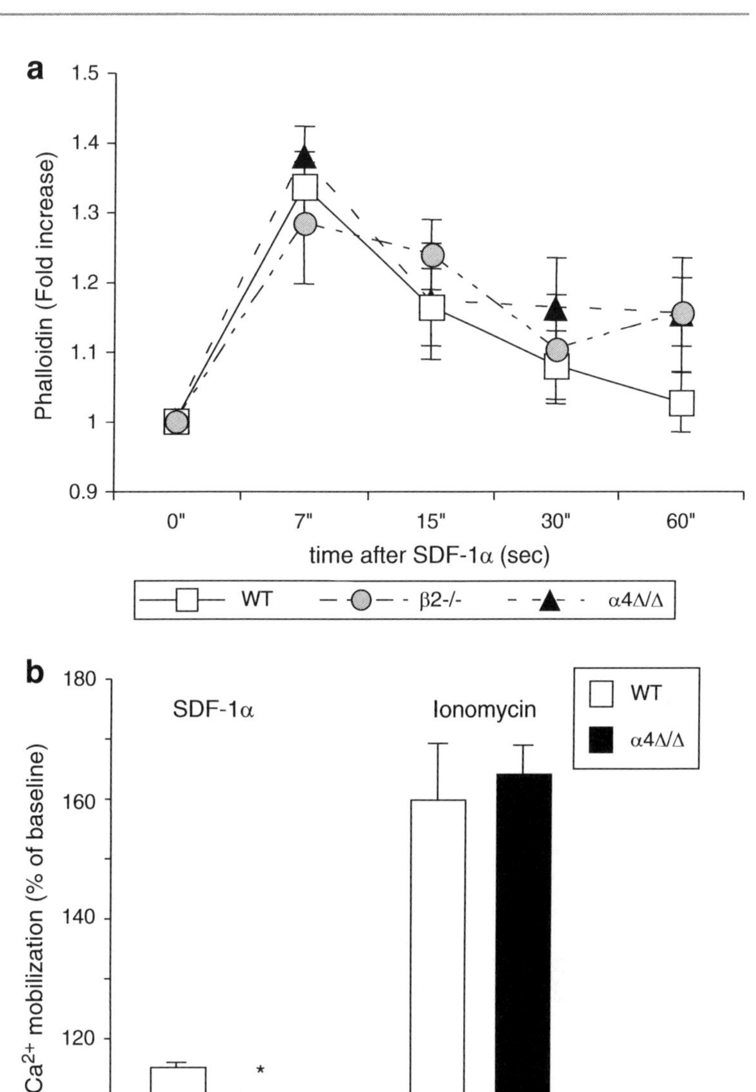

Fig. 6.7 Lymphocyte early responses to chemoattractant stromal-derived factor-1α (SDF-1α). (**a**) Actin polymerization in wild-type (WT), α4Δ/Δ, and $\beta2^{-/-}$ lymphocytes. Splenocytes were stained for lymphocyte markers (CD3 and B220), stimulated with SDF-1α for indicated amount of time and phalloidin staining was performed. Fold increase over nonstimulated cells was calculated. (**b**) SDF-1α–induced Ca^{2+} mobilization. Splenocytes were loaded with Indo-1 and stained for lymphocyte markers. Ca^{2+} flux was assessed following stimulation with SDF-1α or ionomycin. Results are displayed as percent of the peak stimulation levels over baseline. *Significant difference over control ($p < 0.05$). Three mice per group were used

mice at 96 h post-TG injection may suggest either delay in apoptosis, impaired clearance, or sustained neutrophil recruitment. Involvement of both α4 and β2 integrins in regulating apoptosis and/or clearing of apoptotic cells has been demonstrated; deficiency in β2 integrins leads to a delay in neutrophil apoptosis and their subsequent accumulation in the peritoneum [36, 37], while α4 integrin on apoptotic cells assists in their phagocytosis [38]. Thus, absence of α4 integrins may affect clearance of α4Δ/Δ neutrophils, thereby resulting in increased neutrophil presence in the peritoneum.

Lymphocyte migration to the inflamed peritoneum was greatly impaired only in α4 integrin-deficient mice (Fig. 6.4). Our results are in agreement with previously reported observations in chimeric mice in which α4Δ/Δ lymphocytes did not migrate into the inflamed peritoneum [19]. However, the dependence of lymphocyte recruitment on α4 integrin is also a tissue-specific phenomenon; recruitment of lymphocytes to brain tissue is mainly dependent on α4 integrin [39], whereas their migration to inflamed lungs and airways is mediated by both α4 and β2 integrins [11, 40, 41], and migration to inflamed skin depends exclusively upon β2 integrin engagement [42].

It is important to emphasize that, beyond the complete and partial impairment in lymphocyte recruitment seen in α4 and β2-deficient mice, respectively, the defects were manifested differently: α4Δ/Δ lymphocytes showed defective migration to the noninflamed as well as to the inflamed peritoneum; the defect in CD18–/– lymphocytes was partial and became apparent only after inflammation was established (Fig. 6.4a). This finding is in agreement with preferential usage of α4 over β2 integrin by lymphocytes for adhesion to the extracellular matrix before and after inflammatory stimulus [43]. However, whether integrin-specific recruitment changes under shear stress is unclear. For example, preferential usage of β2 integrin by T lymphocytes in response to chemoattractive stimuli under shear stress conditions has been reported [44].

Discussion

Although the lack of either α4 or β2 integrin resulted in decreased efficiency of leukocyte recruitment to the inflamed peritoneum, the substantial numbers of total leukocytes observed at various times after TG injection emphasize the overlapping roles of α4 and β2 integrins in total leukocyte migration. These data are consistent with previous studies showing that only simultaneous functional blockade of α4 and β2 integrins (either by treatment of WT mice with a combination of anti-α4 and anti-β2 antibodies or by treatment of β2-deficient mice with anti-α4-antibody) prevents leukocyte recruitment to the peritoneum [15, 45, 46]. Our data with α4/β2 doubly deficient mice support and extend these observations. In addition, they provide direct evidence for an absence of significant compensation by other cytoadhesins [47, 48] in these animals.

Our studies of the different leukocyte subsets in genetically deficient animals further delineate distinct versus overlapping contributions of α4 and β2 integrins to leukocyte subset-specific migration. Although the presence of either α4 or β2 integrins afforded significant migration of neutrophils to inflamed peritoneum (Fig. 6.1a), assessment of recruitment indices (Fig. 6.1b) indicates that optimal recruitment requires the presence of both α4 and β2 integrins. The interchangeable in CD18_/_ mice [49], the partial defect seen in CD18–/– lymphocytes may also be attributed to an attenuated response to TG of already activated CD18_/_ lymphocytes by preexisting inflammatory stimuli.

The reasons for impaired α4Δ/Δ lymphocyte recruitment to the inflamed peritoneum are attributable mainly to intrinsic cell defects, rather than being dictated by the environment. This is supported by the fact that α4Δ/Δ lymphocytes showed impaired adhesion and migration to SDF-1α (both trans-well and transendothelial migration) (Fig. 6.6b, c). Consistent with these results, perturbed α4 integrin/paxillin-dependent signaling also leads to defective lymphocyte migration to the inflamed peritoneum [26]. Beyond the intrinsic migratory defects of α4Δ/Δ lymphocytes, suboptimal stimulation by environmental factors (i.e., decreased levels of RANTES) cannot be excluded. MMPs that are thought to contribute to leukocyte recruitment [50, 51] most likely do not play a major role in the model employed, as we detected neither lack of induction of either MMP-9 or MMP-2 in the peritoneal cavity during peritonitis nor any difference in transendothelial migration of MMP-9_/_ lymphocytes compared to controls (data not shown). This is in agreement with the observation of Betsuyaku et al. that MMP-9 is not required for efficient leukocyte migration [52].

In summary, our results demonstrate that, although critically dependent on the expression of both α4 and β2 integrins, there is a subset-specific usage of these adhesion molecules by leukocytes in their migration to and egress from the inflamed peritoneum. While the roles of α4 and β2 integrins are partially redundant in myeloid cells, they are distinct and nonoverlapping in lymphocytes. Taken together, our data with genetically deficient mouse models extend and refine previous knowledge on the kinetics of leukocyte accumulation and their egress from the peritoneum in the aseptic peritonitis model. Whether kinetic changes in inflammatory models of other tissues are different will require further studies.

Conclusions

Our data with concurrent use of genetic models of integrin deficiency reveal nonredundant functions of α4 integrins in lymphocyte migration to the peritoneum and further refine specific roles of α4 and β2 integrins concerning trafficking and clearance of other leukocyte subsets at homeostasis and during inflammation.

Acknowledgments This work was supported by National Institutes of Health grants (HL58734, DK46557) for T.P. The author thanks Ms. Devra Batdorf for expert mouse handling.

References

1. Alon R, Feigelson S. From rolling to arrest on blood vessels: leukocyte tap dancing on endothelial integrin ligands and chemokines at subsecond contacts. Semin Immunol. 2002;14:93–104.
2. Yadav R, Larbi KY, Young RE, Nourshargh S. Migration of leukocytes through the vessel wall and beyond. Thromb Haemost. 2003;90:598–606.
3. Chan JR, Hyduk SJ, Cybulsky MI. Chemoattractants induce a rapid and transient upregulation of monocyte alphα4 integrin affinity for vascular cell adhesion molecule 1 which mediates arrest: an early step in the process of emigration. J Exp Med. 2001;193: 1149–58.
4. Kitayama J, Fuhlbrigge RC, Puri KD, Springer TA. P-selectin, L-selectin, and alpha 4 integrin have distinct roles in eosinophil tethering and arrest on vascular endothelial cells under physiological flow conditions. J Immunol. 1997;159:3929–39.
5. Johnston B, Issekutz TB, Kubes P. The alpha 4-integrin supports leukocyte rolling and adhesion in chronically inflamed postcapillary venules in vivo. J Exp Med. 1996;183:1995–2006.
6. Yusuf-Makagiansar H, Anderson ME, Yakovleva TV, Murray JS, Siahaan TJ. Inhibition of LFA-1/ICAM-1 and VLA-4/VCAM-1 as a therapeutic approach to inflammation and autoimmune diseases. Med Res Rev. 2002;22:146–67.
7. Ibbotson GC, Doig C, Kaur J, et al. Functional alphα4-integrin: a newly identified pathway of neutrophil recruitment in critically ill septic patients. Nat Med. 2001;7:465–70.
8. Pereira S, Zhou M, Mocsai A, Lowell C. Resting murine neutrophils express functional alpha 4 integrins that signal through Src family kinases. J Immunol. 2001;166:4115–23.
9. Petruzzelli L, Takami M, Humes HD. Structure and function of cell adhesion molecules. Am J Med. 1999;106:467–76.
10. Mulligan MS, Lentsch AB, Miyasaka M, Ward PA. Cytokine and adhesion molecule requirements for neutrophil recruitment during glycogen-induced peritonitis. Inflamm Res. 1998;47:251–5.
11. Gascoigne MH, Holland K, Page CP, et al. The effect of anti-integrin monoclonal antibodies on antigen-induced pulmonary inflammation in allergic rabbits. Pulm Pharmacol Ther. 2003;16:279–85.
12. Henderson Jr WR, Chi EY, Albert RK, et al. Blockade of CD49d (alphα4 integrin) on intrapulmonary but not circulating leukocytes inhibits airway inflammation and hyperresponsiveness in a mouse model of asthma. J Clin Invest. 1997;100:3083–92.
13. Abraham WM, Sielczak MW, Ahmed A, et al. Alpha 4-integrins mediate antigen-induced late bronchial responses and prolonged airway hyperresponsiveness in sheep. J Clin Invest. 1994;93:776–87.
14. Abraham WM, Gill A, Ahmed A, et al. A small-molecule, tight binding inhibitor of the integrin alpha(4)beta(1) blocks antigen induced airway responses and inflammation in experimental asthma in sheep. Am J Respir Crit Care Med. 2000;162: 603–11.
15. Henderson RB, Lim LH, Tessier PA, et al. The use of lymphocyte function-associated antigen (LFA)-1-deficient mice to determine the role of LFA-1, Mac-1, and alphα4 integrin in the inflammatory response of neutrophils. J Exp Med. 2001;194:219–26.
16. Scharffetter-Kochanek K, Lu H, Norman K, et al. Spontaneous skin ulceration and defective T cell function in CD18 null mice. J Exp Med. 1998;188:119–31.
17. Mizgerd JP, Kubo H, Kutkoski GJ, et al. Neutrophil emigration in the skin, lungs, and peritoneum: different requirements for CD11/CD18 revealed by CD18-deficient mice. J Exp Med. 1997;186:1357–64.
18. Yang JT, Rayburn H, Hynes RO. Cell adhesion events mediated by alpha 4 integrins are essential in placental and cardiac development. Development. 1995;121:549–60.
19. Arroyo AG, Taverna D, Whittaker CA, et al. In vivo roles of integrins during leukocyte development and traffic: insights from the analysis of mice chimeric for alpha 5, alpha v, and alpha 4 integrins. J Immunol. 2000;165:4667–75.
20. Scott LM, Priestley GV, Papayannopoulou T. Deletion of alphα4 integrins from adult hematopoietic cells reveals roles in homeostasis, regeneration, and homing. Mol Cell Biol. 2003;23:9349–60.
21. Vu TH, Shipley JM, Bergers G, et al. MMP-9/gelatinase B is a key regulator of growth plate angiogenesis and apoptosis of hypertrophic chondrocytes. Cell. 1998;93:411–22.

22. Andrew DP, Spellberg JP, Takimoto H, Schmits R, Mak TW, Zukowski MM. Transendothelial migration and trafficking of leukocytes in LFA-1-deficient mice. Eur J Immunol. 1998;28:1959–69.
23. Papayannopoulou T, Priestley GV, Bonig H, Nakamoto B. The role of G-protein signaling in hematopoietic stem/progenitor cell mobilization. Blood. 2003;101:4739–47.
24. Bonig H, Rohmer L, Papayannopoulou T. Long-term functional impairment of hemopoietic progenitor cells engineered to express the S1 catalytic subunit of pertussis toxin. Exp Hematol. 2005;33: 689–98.
25. Geissmann F, Jung S, Littman DR. Blood monocytes consist of two principal subsets with distinct migratory properties. Immunity. 2003;19:71–82.
26. Feral CC, Rose DM, Han J, et al. Blocking the alpha 4 integrin paxillin interaction selectively impairs mononuclear leukocyte recruitment to an inflammatory site. J Clin Invest. 2006;116:715–23.
27. Kuldo JM, Westra J, Asgeirsdottir SA, et al. Differential effects of NF- {kappa}B and p38 MAPK inhibitors and combinations thereof on TNF- {alpha}- and IL-1{beta}-induced proinflammatory status of endothelial cells in vitro. Am J Physiol Cell Physiol. 2005;289:C1229–39.
28. Min JK, Kim YM, Kim SW, et al. TNF-related activation-induced cytokine enhances leukocyte adhesiveness: induction of ICAM-1 and VCAM-1 via TNF receptor-associated factor and protein kinase Cdependent NF-kappaB activation in endothelial cells. J Immunol. 2005;175:531–40.
29. Ding Z, Xiong K, Issekutz TB. Chemokines stimulate human T lymphocyte transendothelial migration to utilize VLA-4 in addition to LFA-1. J Leukoc Biol. 2001;69:458–66.
30. Bowden RA, Ding ZM, Donnachie EM, et al. Role of alphaα4 integrin and VCAM-1 in CD18-independent neutrophil migration across mouse cardiac endothelium. Circ Res. 2002;90:562–9.
31. Burns JA, Issekutz TB, Yagita H, Issekutz AC. The alpha 4 beta 1 (very late antigen (VLA)-4, CD49d/CD29) and alpha 5 beta 1 (VLA-5, CD49e/CD29) integrins mediate beta 2 (CD11/CD18) integrin- independent neutrophil recruitment to endotoxin-induced lung inflammation. J Immunol. 2001;166:4644–9.
32. Mizgerd JP, Bullard DC, Hicks MJ, Beaudet AL, Doerschuk CM. Chronic inflammatory disease alters adhesion molecule requirements for acute neutrophil emigration in mouse skin. J Immunol. 1999;162:5444–8.
33. Hillyer P, Mordelet E, Flynn G, Male D. Chemokines, chemokine receptors and adhesion molecules on different human endothelia: discriminating the tissue-specific functions that affect leucocyte migration. Clin Exp Immunol. 2003;134:431–41.
34. Bellingan GJ, Xu P, Cooksley H, et al. Adhesion molecule-dependent mechanisms regulate the rate of macrophage clearance during the resolution of peritoneal inflammation. J Exp Med. 2002;196: 1515–21.
35. Cao C, Lawrence DA, Strickland DK, Zhang L. A specific role of integrin Mac-1 in accelerated macrophage efflux to the lymphatics. Blood. 2005;106:3234–41.
36. Weinmann P, Scharffetter-Kochanek K, Forlow SB, Peters T, Walzog B. A role for apoptosis in the control of neutrophil homeostasis in the circulation: insights from CD18-deficient mice. Blood. 2003;101: 739–46.
37. Coxon A, Rieu P, Barkalow FJ, et al. A novel role for the beta 2 integrin CD11b/CD18 in neutrophil apoptosis: a homeostatic mechanism in inflammation. Immunity. 1996;5:653–66.
38. Johnson JD, Hess KL, Cook-Mills JM. CD44, alpha(4) integrin, and fucoidin receptor-mediated phagocytosis of apoptotic leukocytes. J Leukoc Biol. 2003;74:810–20.
39. Deloire MS, Touil T, Brochet B, Dousset V, Caille JM, Petry KG. Macrophage brain infiltration in experimental autoimmune encephalomyelitis is not completely compromised by suppressed T-cell invasion: in vivo magnetic resonance imaging illustration in effective anti-VLA-4 antibody treatment. Mult Scler. 2004;10:540–8.
40. Xu B, Wagner N, Pham LN, et al. Lymphocyte homing to bronchus associated lymphoid tissue (BALT) is mediated by L-selectin/PNAd, alphaα4beta1 integrin/VCAM-1, and LFA-1 adhesion pathways. J Exp Med. 2003;197:1255–67.
41. Banerjee ER, Jiang Y, Henderson WR, Scott LM, Papayannopoulou T. α4 and β2 integrins have nonredundant roles for asthma development, but for optimal allergen sensitization only α4 is critical. Exp Hematol. 2007;35:605–17.
42. Grabbe S, Varga G, Beissert S, et al. Beta2 integrins are required for skin homing of primed T cells but not for priming naive T cells. J Clin Invest. 2002;109:183–92.
43. Solpov A, Shenkman B, Vitkovsky Y, et al. Platelets enhance CD4+ lymphocyte adhesion to extracellular matrix under flow conditions: role of platelet aggregation, integrins, and non-integrin receptors. Thromb Haemost. 2006;95:815–21.
44. Schreiber TH, Shinder V, Cain DW, Alon R, Sackstein R. Shear flow-dependent integration of apical and subendothelial chemokines in T-cell transmigration: implications for locomotion and the multistep paradigm. Blood. 2007;109:1381–6.
45. Winn RK, Harlan JM. CD18-independent neutrophil and mononuclear leukocyte emigration into the peritoneum of rabbits. J Clin Invest. 1993;92:1168–73.
46. Henderson RB, Hobbs JA, Mathies M, Hogg N. Rapid recruitment of inflammatory monocytes is independent of neutrophil migration. Blood. 2003;102:328–35.

47. Katayama Y, Hidalgo A, Chang J, Peired A, Frenette PS. CD44 is a physiological E-selectin ligand on neutrophils. J Exp Med. 2005;201: 1183–9.
48. Lenter M, Levinovitz A, Isenmann S, Vestweber D. Monospecific and common glycoprotein ligands for E- and P-selectin on myeloid cells. J Cell Biol. 1994;125:471–81.
49. Wu H, Prince JE, Brayton CF, et al. Host resistance of CD18 knockout mice against systemic infection with Listeria monocytogenes. Infect Immun. 2003;71:5986–93.
50. Kolaczkowska E, Chadzinska M, Scislowska-Czarnecka A, Plytycz B, Opdenakker G, Arnold B. Gelatinase B/matrix metalloproteinase-9 contributes to cellular infiltration in a murine model of zymosan peritonitis. Immunobiology. 2006;211:137–48.
51. Renckens R, Roelofs JJ, Florquin S, et al. Matrix metalloproteinase-9 deficiency impairs host defense against abdominal sepsis. J Immunol. 2006;176:3735–41.
52. Betsuyaku T, Shipley JM, Liu Z, Senior RM. Neutrophil emigration in the lungs, peritoneum, and skin does not require gelatinase B. Am J Respir Cell Mol Biol. 1999;20:1303–9.

Published in

Ulyanova T, Banerjee ER, Priestley GV, Scott LM, Papayannopoulou T. Unique and redundant roles of alphaα4 and beta2 integrins in kinetics of recruitment of lymphoid vs myeloid cell subsets to the inflamed peritoneum revealed by studies of genetically deficient mice. Exp hematol. 2007;35(8):1256–65.

Highlights of the Important Findings from the Critical Analyses of the Data

The facts presented and discussed in the six preceding chapters of this monograph embodies the following important findings from the critical analysis of the data:

I. **On inflammation and inflammatory cell migrates pattern in pulmonary and systemic disorders –**
   1. Functional delineation of small molecule antagonist for more effective drug formulation for treatment of asthma;
   2. Intigrin $\alpha4$ is critical for onset, establishment and maintenance of both acute and chronic allergic asthma, both by regulating sensitization as well as at the tissue remodeling level;
   3. CD18 is key to mechanistic migration of leukocyte across interstitium for the onset of asthma but is redundant for maintenance or exacerbation or structural alteration of airways;
   4. Gp92phox subunit of NADPH oxidase plays a key regulatory role in the onset and development of allergic asthma, both in mature lymphocyte and phagocytes as well as their progenitor cell population in acute asthma.
   5. In chronic asthma, inflammation precedes degeneration and fibrosis and gp91phox acts upstream of MMP-12 to regulate inflammation in phagocytes during onset of the acute phase of idiopathic lung fibrosis.
   6. Gp91phox on macrophages show a functional dichotomy that is strictly tissue specific.
   7. Deletion of three selectins expressed on lymphocytes seems to impede inflammatory migration while only L-selectin also possibly regulates activation of specific T cell subsets in lung and airways.

   Overall, the studies on inflammatory biology of the above disease models reveal hitherto unknown pathways critical for mechanistic regulation of disease etiology and data included in this thesis thoroughly characterizes the same. Identification of these pathways, critical role players and possible cross-talk amongst them delineates them as plausible drug targets for intervention for disease modification at a therapeutic as well as prophylactic level.

II. **On key regulators of lymphopoiesis and lymphoid from primary sites of poiesis –** The studies on lymphomyeloid hematopoietic reconstitution in lethally irradiated Rag 2–/– recipients transplanted with adult $\alpha4$-deficient ($\alpha4\Delta/\Delta$) donor cells explicitly clarifies the role of $\alpha4$ integrins in the repopulation of lymphoid organs , by securing its role in homing to thymus and gut lymphoid tissue and by strengthening previous evidence of attenuated B-cells responses by $\alpha4$- deficient cells. $\alpha4$ integrin–dependent interaction with stromal tissue cells seen to play major roles both homing and intrathymic development of T lymphocytes and for colonization of mucosal tissue by T-lymphocytes cells.

III. **On successful development of an embryonic stem cell-derived induced differentiation of non-ciliated lung lineage specific cells and its validation in successful engraftment and disease amelioration in degenerative pulmonary fibrosis –**
   1. A novel cocktail of nutritive factor preferentially contribute to the development of

E.R. Banerjee, *Perspectives in Inflammation Biology*,
DOI 10.1007/978-81-322-1578-3, © Springer India 2014

a specific cell type over other, that is AEII (alveolar epithelial cell type II) cells grow better in SAGM (small airway growth medium) vs. CC-10+ Clara cells grow better in BEGM (bronchiolar epithelial growth medium);

2. Engraftment of engineered AEII cells validate their functional prowess;
3. Success of engrafted cells in ameliorating disease phenotype and arrest of progression subscribes to the preferred route of administration in the development of a potential cell based therapy in IPF and contributes to the validation of a proof-of-concept to the plausibility and feasibility of such a regenerative therapy.

About the Author

Dr. Ena Ray Banerjee is an Associate Professor of Zoology in University of Calcutta, India, and an alumnus of the premiere educational institutions Lady Brabourne College and Gokhale Memorial Girls' School. She was trained in Immunobiology during her Ph.D. and worked extensively in immune modulation in inflammation in general and cytokine-mediated inflammation in particular. Having taught both under- and post-graduate Zoology as lecturer under University of Calcutta, India, for several years, she pursued her postdoctoral studies as visiting scientist and subsequently faculty of University of Washington, School of Medicine, Seattle, USA. There she began with immunological studies defining key molecules in inflammation and eventually super-specialized into lung inflammation particularly allergy and made a natural transition onto regenerative medicine of the lung, having worked with one of the foremost scientists in stem cell biology, Professor Thalia Papayannopoulou, and a renowned allergist Dr. William R. Henderson, Jr. Her work pioneered tissue engineering of lung lineage specific cells of the non-ciliated variety from human embryonic stem cells and identified stem cell niches in mouse lung. She returned to India and worked for a while in a drug discovery company Advinus Therapeutics, a TATA enterprise, where she led in vitro and in vivo efforts in pharmacological molecules, drug discovery in asthma, and related respiratory illness, and then returned to academics as Reader in her alma mater, the renowned University of Calcutta.

E.R. Banerjee, *Perspectives in Inflammation Biology*,
DOI 10.1007/978-81-322-1578-3, © Springer India 2014

Her group works on drug discovery efforts using novel drugs (small molecules), herbal extracts (functional food), probiotics (nutraceuticals), novel antibody-mediated (camelid antibody), and cells (tissue engineering of stem cells of embryonic origin, adult tissue origin, and umbilical cord derived) in inflammatory disease models (tissue-specific inflammation in the lung and systemic inflammation) and degenerative disease models. She has published widely in premiere scientific journals, and her publications are widely cited in "methods" volumes as well as "drug discovery" web sites and portals. She is also respected as an academician and educationist par excellence and has spearheaded the rejuvenation of a world-class heritage museum because she believes that to do bioprospecting and molecular drug discovery, knowing your biodiversity is the key.

Printed by Publishers' Graphics LLC
LMO131025.15.17.47 20131025